Colli

D

2

Alan Morris and Margaret Baker

ayne de Courcy

D0416317

Published by HarperCollins*Publishers* Limited
77–85 Fulham Palace Road
London W6 8JB

www.**Collins**Education.com
On-line support for schools and colleges

© HarperCollins*Publishers* Ltd 2002

First published 2002

ISBN 0 00 712422 8

British Library Cataloguing in Publication Data
A catalogue record for this book is available from the British Library

Edited by Jane Bryant
Production by Kathryn Botterill
Design by Gecko Ltd
Cover design by Susi Martin-Taylor
Printed and bound by Scotprint

Acknowledgements
The Author and Publishers are grateful to the following for permission to reproduce copyright material:

AEB, a division of AQA (pp. 7, 11, 13, 18, 20, 23, 24, 26, 33, 38, 43, 45, 54, 55, 60, 62, 73, 74). Answers to questions taken from past examination papers are entirely the responsibility of the author and have been neither provided nor approved by the AEB.
Northern Examinations and Assessment Board, a division of AQA (pp. 39, 44, 61, 66, 67, 69). The author is responsible for the suggested answers and the commentaries on the past questions from the Northern Examinations and Assessment Board. They may not constitute the only answers.
OCR (pp. 37, 55). Answers to questions taken from past examination papers are entirely the responsibility of the author and have been neither provided nor approved by the OCR.

Illustrations
Roger Penwill, Gecko Ltd

Every effort has been made to contact the holders of copyright material, but if any have been inadvertently overlooked, the Publishers will be pleased to make the necessary arrangements at the first opportunity.

You might also like to visit:
www.**fire**and**water**.com
The book lover's website

Contents

How this book will help you

by Alan Morris and Margaret Baker

Exam practice – how to answer questions better

This book will help you to improve your A2 Biology grade. It contains lots of questions based on the core topics of the new A2 specifications of AQA, Edexcel, Nuffield and OCR.

You should have two aims in mind in the final stages of preparation for each of your tests or modules – first to maximise your knowledge and understanding of the topics being tested and secondly to ensure that you have good examination technique so that you can score as many marks as possible with the knowledge that you have. This book will help you to improve your examination technique.

Each chapter in this book, except for the last chapter (see the next page), is broken down into four separate elements, aimed at giving you as much guidance and practice as possible:

❶ Exam Question, Student's answer and 'How to score full marks'

The questions and student's answers that we have chosen to start each chapter are typical of the ones you can expect. They show a number of frequent mistakes made by candidates under exam conditions.

The 'How to score full marks' section shows precisely where the student lost marks. In each case we explain how to gain the extra marks so that when you meet these sorts of questions in your exam you will know exactly how to answer them.

❷ 'Don't make these mistakes'

This section highlights the most common mistakes made every year in exam papers. These include poor use and understanding of biological terminology and not reading the question correctly. When you are into your last-minute revision, you can quickly read through all of these sections and make doubly sure that you avoid these mistakes in your exam.

❸ 'Key points to remember'

These key facts pages have been designed to be as concise as possible. Here you will find much of what you need to know in order to answer questions on the core topics in your A2 specification. They are the most important points that you need to cover when revising a particular topic.

❹ Questions to try, Answers and Examiner's comments

Each chapter ends with a number of exam questions for you to answer. Don't cheat. Sit down and answer the questions as if you were in an examination. Try to put into practice all that you have learnt from the previous sections in the chapter. Check your answers through and then look at the answers given at the back of the book. These are answers that would gain full marks and the ticks show where the marks are awarded.

In the 'Examiner's comments' section we highlight where marks are gained and why. Compare your answer with the answer given and decide whether what you have written would gain full marks – and, if not, what aspects you need to improve on.

This book is divided into ten topic-based chapters covering the core Biology and Human Biology topics set by all the exam boards. The chart below lets you see at a glance which topics your particular specification requires you to study at A2 level.

The topics covered by your A2 specification

Topics covered (chapters in this book)	Examination Boards			
	AQA Specification A	AQA Specification B	OCR	Edexcel
Chapter 1 Inheritance	√	√	√	√
Chapter 2 Ecology	√	√	√	√
Chapter 3 Biochemistry of photosynthesis	√	√	√	√
Chapter 4 Biochemistry of respiration	√	√	√	√
Chapter 5 Water transport in plants	√	√	√	√
Chapter 6 Homeostasis	√	√	√	√
Chapter 7 The kidney	√	√	√	√
Chapter 8 Digestion	√	√	√	√
Chapter 9 Nervous system	√	√	√	√
Chapter 10 Transport of respiratory gases	√	√	√	√
Chapter 11 Synoptic questions – essay writing	√	√		√

How A2 Biology exams differ from AS exams

- **More difficult topics are tested** e.g. biochemistry and ecology

- **Synoptic questions are included.** 40% of all A2 questions must be synoptic – this means they have to test links between different areas of the AS and A2 specification. To be able to make links you have to have the basic knowledge in the first place. **Synoptic questions may be set on any topic – and there are many examples in this book.** In Chapter 2 you will find synoptic comprehension questions, and Chapter 11 has lots of synoptic essay questions.

- **There are longer questions – either essays or sections of extended writing.** This is an area where we find candidates are often very weak. For this reason **we have devoted the whole of Chapter 11 to essay writing. We show you how to prepare for and write essays under exam conditions.**

Exam tips

Read the question carefully

Examiners try to make the wording of the questions as straightforward as possible, but in the stress of an examination, it is all too easy to misinterpret a question. **Read every word in every sentence very carefully.** The paper is yours and no examiner will be cross if you underline key words to help you understand exactly what you have to do.

Understand the information

Not all A2 questions are based on recall. Some will be based on unfamiliar material. It is important that you do not panic and are able to **use the principles that you have learnt to answer these types of question.**

- Make sure you identify the area of the specification that the question is about (e.g. 'respiration') – then you can start to think in terms of glycolysis, Krebs cycle, electron/hydrogen carrier system, etc.

- Look at information given to you in a graph or a table very carefully and make sure you give yourself time to **read the information** on the axes or the headings of the table.

- Make sure you understand what this information means before you read the questions.

- Read through longer 'comprehension' passages twice. First, quickly, to identify the topic area and a second time to mark important parts of the passage.

Plan your answer

Each question or part of a question is given a number of marks. Make sure that:

- Your answer gives enough information to get all the marks. For example, if the question is worth 4 marks then you must give 4 points to gain those marks.

- You give enough detail, based on material learnt at A2 level (and AS where relevant).

- All the information given is relevant. Only answer the question set and do not write everything that you know!

Know the meaning of common instructions

Here is a list of some of the common words used in examination questions. Make sure you are familiar with them and that you know what an examiner expects you to do.

- **Describe**

  Replace this word with the word **'what'**. If this term is used in relation to a graph or a table, then you are being asked to recognise a simple trend or pattern within the data and write what it is. You must use the information on the axes or table headings as reference points. 'It goes up and then down' is not enough – what goes up and by how much would be expected at this level.

  In longer questions 'describe' means you need to give a step-by-step account of what is happening. Tell the whole story; a missed step may mean a missed mark.

- **Explain**

  Replace this word with the word **'why'**. This means you must give a biological reason for the pattern you are given in the question. It does not mean the same as describe and although you may need to describe the pattern before you can explain it, no marks will be given if you only describe what is happening.

- **Suggest**

  This is often used when you are not expected to know the answer, but should have enough biological knowledge to put forward a sensible idea.

- **Name**

  This means exactly what it says and requires no more than a one-word answer. You do not have to repeat the question just to put your answer into a sentence e.g. 'The area of the mitochondrion where Krebs cycle occurs is called the matrix.' This only wastes your time.

Inheritance

1 In the flour beetle, the allele for red body colour (R) is dominant to the allele for black body colour (r).

(a) Complete the genetic diagram to show the result of a cross between a heterozygous red beetle and a black beetle.

Parental phenotypes	Red	Black
Parental genotypes	Rr	rr
Gametes	R r	r r
Offspring genotypes	Rr Rr	rr rr
offspring phenotypes	Red Red	Black Black

✓ ①/2

[2 marks]

(b) A mixed culture of red beetles and black beetles was kept in a container in the laboratory under optimal breeding conditions. After one year, there were 149 red beetles and 84 black beetles in the container.

(i) Use the Hardy-Weinberg equation to calculate the expected percentage of heterozygous red beetles in this population.

$p^2 + 2pq + q^2 = 1$ ✓

q^2 is the freq of the black genotype which is **84**   *not a frequency*

q is $\sqrt{84} = 9.1$

$p = 1 - 9.1$

$= -8.1$

$2pq$ is the freq of the heterozygous red beetle

$2 \times 9.1 \times -8.1$

10.1 Answer: 10.1 ①/3

[3 marks]

(ii) Several assumptions are made when using the Hardy-Weinberg equation. Give two of these.

1   Large population ✓

2   No migration ✓   ②/2

[2 marks]

[Total 7 marks]   ④/7

How to score full marks

Follow the rules

Gametes should always be shown in **a circle**. In part **(a)** the student has not clearly shown that the **gametes contain one allele** and did not get a mark. R r could mean that he thinks both alleles are in one gamete. R,r is better. Ⓡ ⓡ is best.

One type of gamete or two?

The genotype of the gametes should identify the **different gametes possible. If only one gamete is possible** – as in part **(a)**: from the genotype rr only r is possible – **this is all that needs to be shown**. Giving both identical gametes will not lose you marks but wastes time and can make the diagram more complicated than necessary.

Calculating frequency

In **(b) (i)** the **number** of beetles **is not the frequency**. The student did remember the Hardy-Weinberg equations but used the **wrong starting point**.

If the **whole population**, i.e. 100%, **shows a characteristic**, the **frequency is 1**. This relationship gives an easy way to calculate frequency.

In this case the total population is:

$149 + 84 = 233$.

As 84 were black the percentage of black beetles is:

$84/233 \times 100 = 36\%$

The frequency is thus 0.36.

Remember the easy answers

There are a number of factors that affect the Hardy-Weinberg equation, including the fact that **mating must be random** and that **all gametes must be equally viable**. But why make life complicated? If there are five or six different options remember two or three of the easy ones (as the student did in part **(b) (ii)**) – you are unlikely to be asked for a huge list.

Don't make these mistakes...

Although **every body cell contains two alleles** (types of each gene), every **gamete only has one allele**. As each allele is represented by a letter, when you are asked for the genotypes in questions remember:

a **body cell** will be represented by **two letters**

a **gamete** will be represented by **one letter**.

Remember:
- the frequency of the **alleles** is represented by **p** or **q**
- the frequency of the **homozygous genotype** is represented by p^2 or q^2

Do not reverse them!

Never waste valuable time by writing out the question. The minimum amount of space that can be given for your answer is a single line; don't feel you have to fill it. Most of the exam questions are 'structured', so the question is right above your answer. **A single correct word is sometimes all that is expected.**

Work out in advance how much time you can spend to get each mark – it is normally about one and a half minutes per mark. Some questions have a lot more things to read or diagrams to look at. This can be time consuming, but be sure that you **have all the information** and all the **clues** the examiner is offering **before you begin** your answer.

Key points to remember

Terms and conventions

- The genetic material that codes for a particular characteristic is a **gene**.
- There may be **more than one type** of that gene.
- Each type of a gene is called an **allele**.
- There may be more than two types of an allele – **multiple alleles**.
- **Body cells** contain only **two alleles** of each gene.
- **Gametes** – sex cells – contain only **one allele** of every gene.
- Some alleles express their phenotypic character with only one example in a body cell – they are **dominant alleles**.
- Some alleles will be expressed only if two are present in a body cell – they are **recessive alleles**.
- If **both** alleles are the **same**, the genotype is **homozygous**.
- If the **alleles are different**, the genotype is **heterozygous**.
- **Dominant** alleles are represented by a **capital** letter (such as A).
- **Recessive** alleles are represented by a **lower-case** letter (such as a).
- **Multiple alleles** are represented by a **capital** letter with **superscripts** for each allele (e.g. I^A, I^B, I^o). (Capital and lower-case letter still represent dominant and recessive relationships between the alleles.)

Monohybrid cross

- **Mono** refers to a **single gene**; **hybrid** refers to the genotype having **one of each allele**.
- Therefore if gene A is the single gene:

    A and a are two alleles of that gene.

    The cross is between two individuals, both with the genotype Aa.

    Presuming that one allele is dominant and the other recessive the phenotypic ratio will be 3:1.

Dihybrid cross

- **Di** refers to **two genes** being involved; **hybrid** refers to the genotype having **one of each allele**.
- Therefore if genes A and B are the two genes:

    A and a are two alleles of gene A

    B and b are two alleles of gene B

    The cross is between two individuals, both with the genotype AaBb.

    Presuming that one allele is dominant and the other recessive, in both cases the phenotypic ratio will be 9 : 3 : 3 : 1.

Layout

Genetic crosses must always be laid out in this way:

Phenotype of parents
You are often given this in a question.

Genotype of parents
You may be given the letters to use – if not, **use one letter to represent each gene** (try to avoid letters which have capitals and lower-case letters that can easily be confused like c or s). The information in the question will allow you to work out whether the genotype is homozygous recessive, homozygous dominant or heterozygous.

Genotype of gametes
If the genotype is AA there is no need to work with two identical gametes. **Always put the letter used to represent each gamete in a circle.**

Genotype of offspring
There is an equal chance of each male gamete meeting each female gamete. Draw a table to show all possible crosses.

Phenotype of offspring
Use your knowledge of the dominant or recessive nature of the alleles to work out each possible phenotype.

Ratio
Calculate the number of each phenotype.

Key points to remember

Meiosis

- **At the beginning** of both mitosis and meiosis, chromosomes **consist of pairs of chromatids**.

- Chromosomes behave independently during mitosis but during the first stage of meiosis **homologous chromosomes associate** (i.e. move close to each other/pair up).

- Genetic material may be exchanged between homologous chromosomes – **cross-over**.

- During the **first division** of meiosis the homologous chromosomes **separate** – but each chromosome is still made up of **two chromatids**.

- The **separation** of each type of chromosome does not affect the separation of any other – **independent assortment**.

- During the second division of meiosis and the only division of mitosis these chromatids separate.

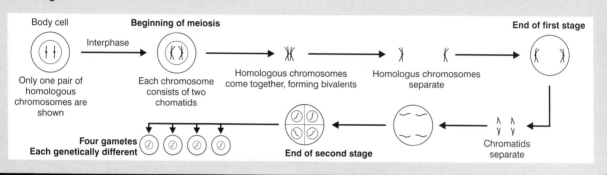

Hardy-Weinberg

- The frequencies of genes within a population are impossible to see. This data may be important for tracing the development of certain conditions or to calculate the chance of someone developing a condition. We can only count the numbers of individuals with a particular phenotype. Knowing the total population we can then calculate the frequency of individuals with this phenotype. The Hardy–Weinberg equation enables us to calculate the genotype and allele frequencies

- Provided that certain conditions exist the **frequency of alleles** in a population will remain the **same over generations** – so will the frequencies of the **genotypes** and the **phenotypes**.

- The conditions are: a **large population**, which is **isolated** from other populations; **random mating** (but only within that generation); all **gametes** are **equally viable**; all **genotypes** are **equally viable**; **no mutation**.

- There are two equations that relate the frequencies of alleles of a gene and the frequencies of the genotypes of the same alleles.

  Using the alleles R and r as examples:

  If p = frequency of allele R (dominant allele) and q = frequency of allele r (the recessive allele)

then

$$p + q = 1$$

If p^2 = frequency of genotype RR (the homozygous dominant) and q^2 = frequency of genotype rr (the homozygous recessive) and $2pq$ = frequency of genotype Rr (the heterozygous)

then

$$p^2 + 2pq + q^2 = 1$$

- The only frequency you can be sure of (unless you are given it in the question) is that of the homozygous recessive genotype, q^2.

  However, from that you can work out everything else!!

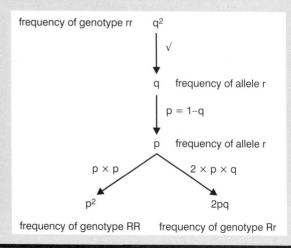

Questions to try

Q1

The diagram shows the inheritance of ABO blood groups in a family.

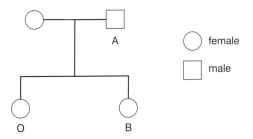

(a) What is:

(i) The phenotype of the mother;

....

[1 mark]

(ii) the probability that the next child will be a girl of blood group AB?

....

[1 mark]

(b) In the population of Iceland, the frequency of the I^A allele is 0.19 and the frequency of the I^B allele is 0.06. Showing your working in each case, calculate:

(i) the frequency of the I^O allele;

....

....

[1 mark]

(ii) the expected percentage of the Icelandic population with the genotype $I^O I^O$.

....

....

[2 marks]

[Total 5 marks]

Examiner's hints
- Only during meiosis will the exchange of genetic material take place between different chromosomes.
- During the first stage of meiosis homologous chromosomes separate.
- Only during meiosis will paired structures (shaped like double Vs) be seen moving to the poles.
- During the second stage of meiosis chromatids separate

Q2

The diagram shows some of the stages in one type of cell division. Only one pair of homologous chromosomes is shown.

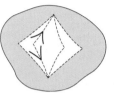   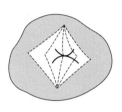

    A   B   C   D

(a) Giving one reason, identify the type of cell division.

....

....

[2 marks]

(b) Place stages A, B, C and D in the correct order.

....

[1 mark]

(c) Explain the importance of the events occurring in stage C.

....

....

[2 marks]

[Total 5 marks]

Answers to Questions to try are on pages 85–86.

2 Ecology

Exam question and student's answer

Read the following passage

> *Azolla pinnata*, the world's smallest fern, is used as a fertiliser in rice paddies because it is so rich in nitrogen – a nutrient often in short supply in the environment. The nitrogen is concentrated in the fern by the activities of a <u>photosynthetic bacterium</u> called *Anabaena* which lives within its leaves. The bacterium takes <u>nitrogen from the air</u> (which exists as N_2 – a form not usable by plants), splits it, and then combines it with hydrogen to produce <u>ammonium ions</u> (NH_4^+). Nitrogen in this form can be absorbed and used by the plant.
>
> *nitrogen fixation*
>
> Ammonium synthesis takes place in specialised cells called heterocysts. *Anabaena* can easily be seen with a light microscope and looks like a <u>string of beads</u>, with each bead representing a blue-green photosynthetic cell. The heterocysts stand out among these as occasional, large, colourless, <u>thick-walled cells</u>. The thick walls of the heterocysts exclude oxygen which would otherwise inactivate the enzymes of ammonium synthesis inside the cell.
>
> *structure of bacteria*
>
> *Anabaena* contains two enzymes, reductase and nitrogenase, which catalyse the conversion of N_2 to NH_3. Reductase transfers electrons from a donor molecule to nitrogenase which reduces N_2 to NH_3. <u>Hydrolysis of ATP to ADP triggers this transfer of electrons</u>. Once formed the NH_3 (which in aqueous solution exists predominantly as the NH_4^+ ion) is released into the soil. The *Azolla* plants growing in the paddy fields eventually die and decompose, releasing yet more NH_4^+ ions into the soil. These ions may be converted into nitrites and nitrates by other soil microorganisms.
>
> *biochemistry and enzymes*
>
> Nitrate ions are absorbed by rice plants where the nitrogen is incorporated into amino acids and the bases of nucleic acids. And so the chemically resistant nitrogen finds its way into living organisms. The result is more and larger rice grains with no need for chemical fertilisers.
>
> *assimilation*
>
> <u>Azolla</u> is not just used for fertilising rice. It can be fed to cattle, pigs, ducks, chickens and carp which can then be <u>eaten by humans</u>. Presently, rice covers nearly 11% of the world's arable land. Given our ever-increasing population, humans will need more rice and would do well to depend on a fertiliser which is as environmentally safe and energy-efficient as *Azolla*.
>
> *food chains*

(Line numbers: 5, 10, 15, 20)

Source: adapted from *Biological Sciences Review*, November 1997

Using information from the passage and your own knowledge, answer the following questions.

(a) The passage describes the structure of the nitrogen fixing microorganism *Anabaena*. Draw a simple diagram to show the appearance of *Anabaena* when viewed with a light microscope. Indicate on your diagram the sites of photosynthesis and nitrogen fixation.

Photosynthesis ✓   Nitrogen fixation ✓   [2 marks]   2/2

(b) **(i)** Which lines in the passage describe the process of nitrogen fixation?

Lines <u>18 and 19</u>   [1 mark]   0/1

(ii) What is the evidence in the passage that shows that oxygen gas could be a hindrance to nitrogen fixation?

The special cells where nitrogen fixation occur have thicker walls to stop oxygen getting in.   [1 mark]   0/1

(c) The relationship between electron-transfer and the ATP/ADP interconversion is described in lines 12–14. In what way is the relationship described here different from that occurring in a mitochondrion during aerobic respiration?

ATP is broken down to ADP but in the mitochondria ADP combines with phosphate to make ✓ATP.

(1/1)

[1 mark]

(d) Describe the roles of decomposition and nitrification in the transfer of nitrogen-containing compounds from *Azolla* to rice plants.

Decomposition breaks down Azolla into NH_4^+ ✓ions in the soil. These are converted into nitrates ✓by Nitrosomonas during nitrification.

(2/3)

[3 marks]

(e) Using only information from the passage, give two simple food chains, based upon the productivity of Azolla:

(i) showing humans as primary consumers;

Azolla → human

(0/1)

[1 mark]

(ii) showing humans as secondary consumers.

Azolla → cattle → human ✓

(1/1)

[1 mark]

[Total 10 marks] (6/10)

How to score full marks

This is a typical example of a comprehension question. Although the material could come from any area of the specification, the advice given here can be applied to any comprehension question.

Before you start

In a comprehension you need to **relate the topic to what you know.** Food chains and the nitrogen cycle are popular topics, but often cause students problems. Review what you know – ammonification, nitrification, denitrification, nitrogen fixation. How do they fit in with the passage and the questions?

Use the information

In part **(a)** there is a clear description of the bacterial cell's structure so a drawing is easy. **Do not forget the instructions** to label the sites of photosynthesis and nitrogen fixation. **Read the passage**: it is clear what happens where.

🎯 Don't get confused

One of the biggest areas of confusion is between the processes of nitrification and nitrogen fixation. In part **(b) (i)** the student has identified nitrification, rather than nitrogen fixation, which is described in lines 4–5 or 13–14.

🎯 Take words from the passage

Most comprehension passages will contain statements that examiners expect you to use in your answers. In part **(b) (ii)** the words 'what evidence from the passage' tells the student to quote terms such as 'heterocyst', which he failed to do.

🎯 Look at the number of marks you need to score

The mark allocation gives an indication of the number of points you need to make or the stages of your argument. In part **(d)** the passage described the transfer of nitrogen atoms through different compounds – ammonium ions, nitrites and nitrates. The student, however, jumped from ammonium to nitrate. **Be sure that you 'tell the whole story'.** The student gave the name of one of the bacteria involved and thought he had made a point for the third mark. Remember that the names of the bacteria involved in the nitrogen cycle are not required by most specifications.

🎯 Know your principles and read the passage carefully

A primary consumer is a herbivore, something that eats a producer (a plant). The student correctly described the food chain involving humans as a secondary consumer (a carnivore) in part **(e) (ii)**. However, he missed the fact that *Azolla* is used to fertilise rice (a plant and so a producer) eaten by humans (being herbivores now, so a primary consumer). He thus tried to link the bacteria to humans directly, which is not true.

Don't forget ... Comprehension questions

The following comments will help you cope more easily with this type of synoptic question.

Comprehension passages give examiners the chance to write synoptic questions – quite often they **link biological principles and their applications.**

Long passages contain lots of information, much more than you will need. **Do not try to remember it all.** Do not be scared of long or unfamiliar words.

Use the passage when you can. Quote words or phrases when you get the chance, but remember this material could be used as a way of introducing some related topic not actually in the passage.

Read the passages all the way through, to get an idea of the topic, before you read the questions.

Re-read the passage more slowly. **Highlight or underline important phrases. Summarise the major areas** dealt with in each paragraph. You might think that this is a waste of time but time has been allocated to allow you to do this.

Comprehension passages can be taken from scientific journals like the 'New Scientist' or 'Biological Review' or from newspaper articles. Look back at relevant topics and past paper questions and **be aware of the sort of passage that can be used.**

Read biological material from sources other than your textbook such as scientific articles from newspapers. There is no need to go overboard and to subscribe to lots of journals, simply **see the different ways facts can be written.**

Key points to remember

Key terms

- **Population** — A group of organisms of the **same species** living in the **same area**.
- **Community** — **All the populations** living in a particular place.
- **Abiotic factors** — **Non-living** factors that may influence a population, such as temperature.
- **Biotic factors** — **Living** factors that may influence a population – such as an animal that might eat another animal living in the same area.
- **Ecosystem** — The **interaction** of all the abiotic and biotic factors.
- **Habitat** — Where the organism lives (its address).
- **Niche** — What the organism does in the ecosystem (its profession).
- **Intraspecific competition** — Organisms compete with their own species for factors like food.
- **Interspecific competition** — Organisms compete with other species for factors like food.

The carbon cycle

- The only way producers can take up carbon is by photosynthesis using carbon dioxide.
- Carbon dioxide is released by all living organisms as a waste product of respiration.

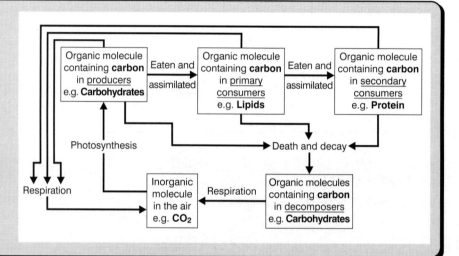

The nitrogen cycle

- Nitrogen is available to decomposers when organisms die (as protein) or through excretory products (as urea).
- Two steps are involved in converting organic nitrogen-containing compounds into molecules that can be absorbed:
  1 ammonification.
  2 nitrification.
- Nitrogen fixation incorporates nitrogen gas into nitrogen-containing compounds.
- Denitrification produces nitrogen gas from nitrogen-containing compounds.

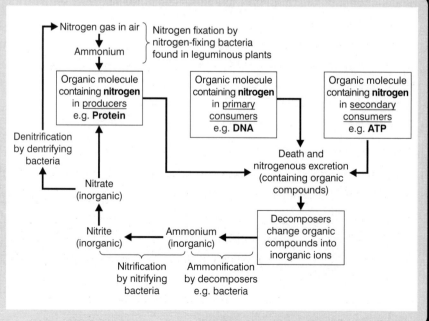

Ecological pyramids

- Food webs show only qualitative information.
- Pyramids of numbers, biomass and energy give **quantitative** information.

Different ecological pyramids

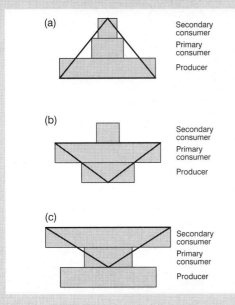

Pyramids of number

- Allow us to compare the number of organisms in each trophic level at a particular time.
- Usually pyramid-shaped (Figure (a)), but two important exceptions result in inverted pyramids (b and c):
  1 a lot of small animals feeding on a large plant (e.g. many aphids on a rose bush;(b))
  2 an animal with a large number of small parasites feeding on it (e.g. one dog with many fleas;(c)).

Pyramids of biomass

- Allow us to compare the total amount (mass) of living material present at each trophic level, at a particular time.
- Overcome the problem associated with differences in the size of organisms.
- Most are pyramid-shaped, but one important exception gives an inverted pyramid (Figure b): When the producer is a small organism which multiplies very rapidly the total biomass of the producer at **any one time** may be less than the total biomass of the primary consumer.

Pyramids of energy

- Allow us to compare the amount of energy at each trophic level over a period of time.
- Pyramids of energy are **always** pyramid-shaped.

Energy transfer

By calculating the energy taken into an organism and that converted into biomass, we can work out the efficiency of the energy transfer from one trophic level to the next.

Energy transfer through the food chain

(a) Not all the light energy falling on a plant makes new tissue

- Some is of the wrong wavelength for photosynthesis.
- Some does not strike chlorophyll.
- Some will be reflected from the plant surface.

(b) Not all the chemical energy taken in by the consumer is used to make new tissue

- Some is lost in faeces.
- Some is used in respiration.

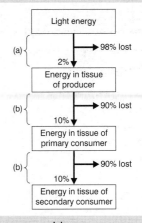

Efficiency of transfer

The efficiency of energy transfer between trophic levels is about 10%, but is very variable.

- The diet of herbivores contains more cellulose, so more energy is lost in faeces.
- Some secondary consumers are more active than primary consumers, and lose more energy in respiration to provide the ATP for movement.
- Poikilothermic animals (e.g. reptiles, fish) use less energy producing body heat than homoeothermic animals (e.g. birds, mammals).
- Small mammals have a large surface area to volume ratio, and will lose more body heat.

Questions to try

Q1

The diagram shows a nitrogen cycle for an arable farm in which all parts of the crop above ground are removed at harvest. The figures are in kg of nitrogen ha^{-1} year^{-1}.

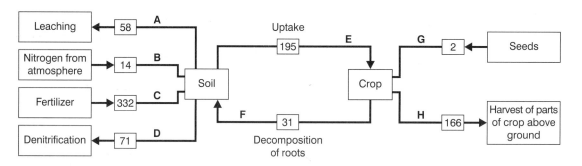

(a) Calculate the net annual gain in soil nitrogen.

....

[1 mark]

(b) Give the letter of one pathway involving:

(i) nitrifying bacteria; ....

[1 mark]

(ii) nitrogen-fixing bacteria. ....

[1 mark]

(c) **(i)** Describe what happens in the process of denitrification.

....

[1 mark]

(ii) This farm has heavy clay soil which easily becomes waterlogged. Explain why the figure for denitrification would be lower on a farm with sandy soil.

....

....

[2 marks]

[Total 6 marks]

Q2

The diagram shows the fate of food energy consumed by a sheep. All units are MJ day^{-1}.

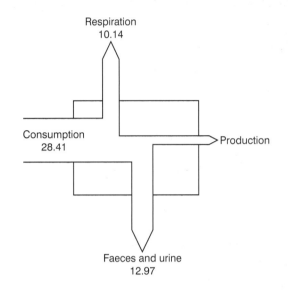

(a) What will happen to the energy contained in faeces and urine?

....

....

....

[1 mark]

(b) Calculate:

(i) the amount of energy that goes into the production of new tissue by the sheep;

Answer = ...

[1 mark]

(ii) the efficiency of energy transfer by this animal.

Answer = ...

[1 mark]

(c) Intensive farming involves keeping animals inside and feeding them as required. Giving an explanation for your answer, suggest one way in which intensive farming can improve the efficiency of energy transfer.

....

....

....

[2 marks]

[Total 5 marks]

Answers to Questions to try are on pages 86–87.

Exam question and student's answer

(a) The diagram summarises the light-independent reaction of photosynthesis.

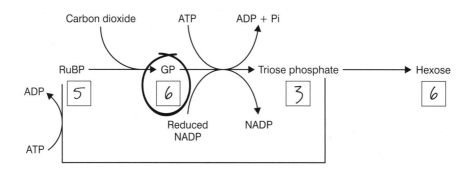

(i) Complete the four boxes to show the number of carbon atoms in a molecule of each substance.

[1 mark]   0/1

(ii) Where in the chloroplast does the light-independent reaction take place?

The light-independent reaction takes place in the stroma. ✓

[1 mark]   1/1

(iii) Explain why the amount of GP increases after a photosynthesising plant has been in darkness for a short time.

In the dark no ATP or reduced ✓ NADP can be formed

because that needs light. These things will be formed in

the light-dependent reaction.

[2 marks] 1/2

(b) Describe the role of water in the light-dependent reaction of photosynthesis.

Light excites chlorophyll and an electron is removed. This is

picked up by a series of carriers and ATP is made. Chlorophyll

receives an electron ✓ from water which causes it to break down,

this process is called photolysis. Hydrogen released from water is

picked up by NADP making reduced ✓ NADP. ATP and reduced

NADP are used in the light independent reaction.

[2 marks] 2/2

[Total 6 marks]   4/6

How to score full marks

Things you must learn

Part (a) (i) expects you to **remember the number of carbon atoms** in certain molecules. Many are obvious – **TRI**ose: **3** or **HEX**ose: **6**. The student's mistake was to forget that when a 5 carbon (RuBP) and a 1 carbon (carbon dioxide) molecule combine the molecule formed is so unstable that it breaks down to form two molecules of a 3-carbon compound, GP.

Look at the whole picture

When information is given in the form of a diagram **look carefully at the diagram** to see what you need to use. In part (a) (iii) the student did realise that no light would mean no ATP or reduced NADP but there is a logical series of steps that follows as a result of this:

- GP can not be changed to TP.
- TP can still be changed to hexose.
- So the level of TP goes down.
- RuBP can still be changed to GP.
- So the level of GP goes up.
- TP can not be changed to RuBP (no ATP – see the diagram).
- So the level of RuBP goes down.

It is possible to ask about the change of level of any of these molecules or to give a graph showing the changes and ask for an explanation.
Be prepared for any variation on this theme.

Read the question

Be sure you know where the question wants you to start! In part (b) water's role is to replace the electron lost from chlorophyll and to provide the proton (hydrogen ion, H^+) to reduce NADP. **What happens before and after is irrelevant** and will not get you any marks.

Be aware of links

Before revising this topic look back at the biochemical sections of your AS specifications. There are many examples of synoptic links here – e.g. carbohydrate structure, enzymes.

Don't make these mistakes...

If you are asked 'where' something is, there is no need to put your answer into a sentence. Doing so will not lose you marks, but it will not gain any either – and it will waste time.

Don't write too much. The space in the exam paper allows plenty of room for a complete answer – even if you have large handwriting! If you are in the habit of writing lots more – think again. Are you 'writing all you know about ...' rather than answering the question?

Key points to remember

Basic word equation for photosynthesis:

$$\text{Carbon dioxide} + \text{Water} \xrightarrow[\text{Chlorophyll}]{\text{Light energy}} \text{Carbohydrate} + \text{Oxygen}$$

- This only gives the raw materials and end products.
- Photosynthesis consists of **two series of reactions**:

  light dependent, followed by

  light independent.

The light-independent reaction

- Carbon dioxide combines with ribulose bisphosphate (RuBP).
- Two molecules of glycerol-3-phosphate (GP) are formed.
- GP is reduced to triose phosphate (TP).
- Reduced NADP (from the light-dependent reaction) provides the hydrogen.
- ATP (from the light-dependent reaction) provides the energy.
- ATP (from the light-dependent reaction) provides the phosphate to reform RuBP.
- The **end products** of the light-independent reaction are **organic molecules**.

The light-dependent reaction

- Chlorophyll is photosensitive.
- **Light excites** one of the **electrons** of chlorophyll, which gains so much energy that it **leaves** chlorophyll and is picked up by a **carrier**.
- This electron is **passed** from carrier to carrier.
- As the electron is passed from one carrier to another **energy is released** and is used to combine ADP and phosphate to form ATP.
- This is called **Photophosphorylation**.
- Water is broken down – **photolysis** – releasing oxygen, electrons (e^-) and protons (H^+).
- **Oxygen** is released as a **waste product**.
- The electrons replace those lost by chlorophyll.
- The protons from water and the electrons from chlorophyll reduce the hydrogen acceptor NADP, forming **reduced NADP**.
- The **end products** of the light-dependent reaction are **ATP** and **reduced NADP**.

Summary of photosynthesis

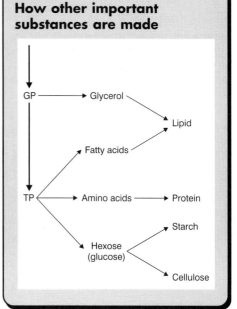

How other important substances are made

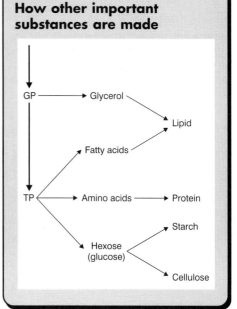

Questions to try

Q1

The diagram summarises the light-dependent reaction in photosynthesis.

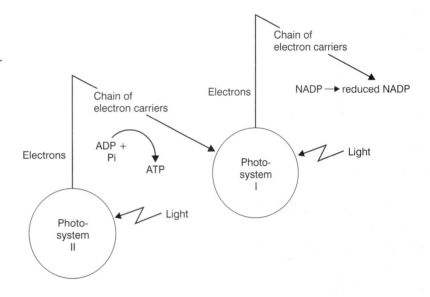

(a) Where, in the chloroplast, does the light-dependent reaction take place?

....

[1 mark]

(b) During this reaction water molecules are broken down to yield oxygen, electrons and hydrogen ions (protons).

(i) What is the name given to the process in which the water molecules are broken down?

....

[1 mark]

(ii) What happens to the electrons produced in this process?

....

....

[1 mark]

(iii) What happens to the hydrogen ions?

....

....

[1 mark]

[Total 4 marks]

Examiner's hints

● The light-dependent reactions take place in the chloroplast membranes or thylakoids. It requires light energy, water, NADP, ADP and phosphate.

● The light-independent reactions take place in the area surrounding the membranes, the stroma. It requires carbon dioxide, RuBP, and the products of the light-dependent reaction; ATP and reduced NADP.

● Both reactions occur in complete chloroplasts.
     Only the light-dependent reactions occur if the membranes are isolated
     Only the light-independent reactions if the stroma is isolated.

Q2

In an investigation to find the precise location of the light-independent reaction of photosynthesis, the procedure shown in the diagram was followed.

(a) In Stage 1 of the procedure:

   (i)   describe what happens to the ADP and NADP;

   ....

   ....

   [2 marks]

   (ii) explain why it was necessary to exclude carbon dioxide.

   ....

   ....

   [2 marks]

(b) Suggest how the contents of the tube might have been analysed in Stage 3 to show that ^{14}C was now in organic compounds.

   ....

   ....

   [2 marks]

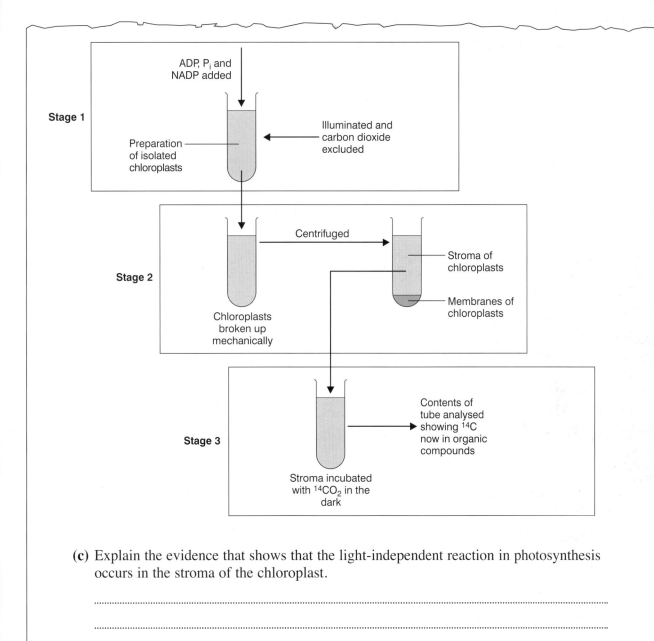

(c) Explain the evidence that shows that the light-independent reaction in photosynthesis occurs in the stroma of the chloroplast.

....

....

[2 marks]

[Total 8 marks]

Answers to Questions to try are on page 87.

Exam question and student's answer

The diagram summarises some of the steps in the biochemical pathway of respiration. The names in italics refer to the enzymes which catalyse certain stages.

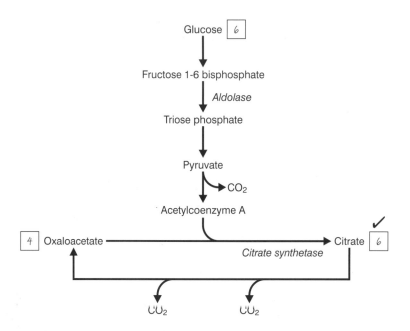

(a) Complete the diagram by writing the number of carbon atoms contained in the compounds or ions in the appropriate boxes.

[1 mark] ①/₁

(b) The enzyme citrate synthetase is inhibited by ATP.

(i) ATP acts as a non-competitive inhibitor. Explain how ATP inhibits the production of citrate.

ATP does not compete with oxaloacetate for the ∧active site ✓ but it does change the shape of the active site and therefore stops the production of citrate.

[3 marks] ①/₃

(ii) The amount of ATP produced by a cell is regulated by negative feedback. Explain how the inhibition of citrate synthetase by ATP is an example of negative feedback.

An increase in the amount of ATP reduces ✓ the activity of the enzyme. This in turn slows down Krebs ✓ cycle which reduces the amount of ATP being produced.

[2 marks] ②/₂

In an investigation, ultracentrifugation was used to separate the components of liver cells. The flow chart summarises the steps in the process.

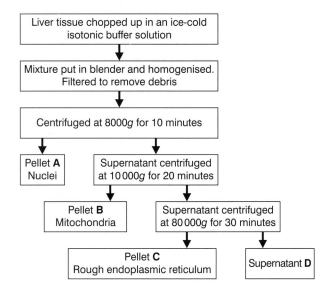

(c) Explain why it was necessary to chop up the liver in a buffer solution.

A buffer solution will either add hydrogen ions or absorb them, in this way the pH of the liver will always be the same. ✓ ∧

1/2

[2 marks]

(d) Explaining your answer in each case, which of pellets **A**, **B** or **C** would be associated with:

(i) the highest concentration of DNA;

A, ✓ because DNA is a component of chromosomes ✓ which are found in the nucleus.

2/2

[2 marks]

(ii) polymerisation of amino acids?

C, ✓ because amino acids are joined together in the ribosomes ✓ which are on the rough ER.

2/2

[2 marks]

(e) Describe a biochemical test that would enable you to demonstrate that supernatant D contained a protein.

Add <u>Iodine solution</u> and the supernatant will go purple. ✗

0/2

[2 marks]

[Total 14 marks]

9/14

How to score full marks

Remember the principles

If principles learnt at AS are being revisited, as in part **(b)** – enzyme inhibition and homeostasis – **tell the whole story**. Although the student did recall that homeostasis involves changes in the variable itself (in this case an increase of ATP) causing a return to the norm (stopping any more ATP being made), she was rather less convincing when she described enzyme inhibition. A non-competitive inhibitor combines with the enzyme but **not at the active site**. This, however, changes the shape of the active site and therefore makes it impossible for the substrate to form an enzyme/substrate complex.

Don't spot words

In part **(c)** the student has spotted the word buffer and has described how a buffer works. She has, luckily, said that this will maintain a constant pH but missed the importance of this in terms of the experimental procedure. Again **this is synoptic** – enzymes are affected (and you should be able to remember why) by changing pH, so a buffer will stop that happening.

Put things into your own words

Sometimes the language used in exam papers makes it hard to see what is needed. If this happens **put phrases into your own words.** Part **(d) (ii)** is an example: 'polymerisation of amino acids' – polymerisation means joining lots of the same molecule together; those molecules are amino acids. What do you get when you put lots of amino acids together? A protein. So this phrase means 'protein synthesis'. Where does protein synthesis happen? The ribosome; an organelle attached to the endoplasmic reticulum – so pellet **C**.

Know your reagents and colour changes

Food tests are simple recall. **You must know what reagent to use** to test for the common types of organic molecule – protein, lipid, starch, reducing sugar, non-reducing sugar. In part **(e)** Biuret reagent should have been used, which would go a pale purple in the presence of protein. Iodine solution goes a similar colour in the presence of starch, so the student lost two marks here, simply for failing to remember the right reagent.

Don't make these mistakes...

Remember the enzymes. As every step of respiration is controlled by an enzyme, this is another synoptic topic which can be tested. Make sure you know how enzymes work, what factors affect them and how they can be inhibited.

Remember where each of the stages of respiration takes place:

Glycolysis – cytoplasm

Link reaction – matrix of the mitochondria

Krebs cycle – matrix of the mitochondria

Electron transport chain – cristae of mitochondria

Don't forget the key molecules in respiration. You do not need to know the names of all the molecules involved in respiration but remember these.

Molecule	Number of carbon atoms
Glucose	6
Triose phosphate (TP)	3
Glycerate phosphate (GP)	3
Pyruvate	3
Acetyl coenzyme A	2
Oxaloacetete	4
Citrate	6

Key points to remember

ATP

- ATP is hydrolysed to ADP and a phosphate group (Pi). This releases energy for reactions in the cell.
- ATP must be continually reformed from ADP and phosphate. This reaction needs a source of energy.

Energy from reactions such as respiration

$$ADP + P_i \rightleftharpoons ATP$$

Energy transferred to reactions requiring energy

Overview of respiration

- Respiration is often represented by the equation:

$$C_6H_{12}O_6 + 6O_2 \rightarrow 6CO_2 + 6H_2O + Energy$$

- $C_6H_{12}O_6$ represents glucose, but lipids and protein can also be used.
- Oxygen is not always necessary for respiration.
- Respiration always takes place in cells and results in the release of energy.

Aerobic respiration

There are three main stages.

1 Glycolysis

- A molecule of glucose is broken down into two molecules of pyruvate.

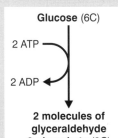

Glucose (6C)

2 ATP → 2 ADP

*The ATP supplies the phosphate group to add to the glucose.
So here 2 molecules of ATP are used.*

2 molecules of glyceraldehyde 3-phosphate (3C)

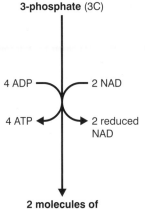

4 ADP → 4 ATP

2 NAD → 2 reduced NAD

*This is now broken down and in the process produces 4 ATP molecules, so there is a **net gain of 2 ATP** for each molecule of glucose. Also the reactions release hydrogen. This is removed and used to reduce a coenzyme, NAD, forming **reduced NAD**. The reduced NAD will be used in a later stage.*

2 molecules of pyruvate (3C)

2 The link reaction

- The reactions linking glycolysis to the Krebs cycle.

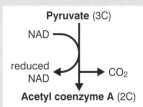

Pyruvate (3C)

NAD → reduced NAD → CO_2

Acetyl coenzyme A (2C)

The reaction releases hydrogen, which is used to reduce NAD. This reaction involves the loss of CO_2.

3 Krebs cycle

- A series of reactions which produces ATP, releases hydrogen (which reduces coenzymes NAD and FAD), releases CO_2.

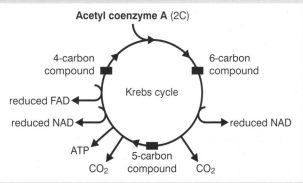

Acetyl coenzyme A (2C)

4-carbon compound

6-carbon compound

reduced FAD

reduced NAD

Krebs cycle

reduced NAD

ATP

CO_2

5-carbon compound

CO_2

Hydrogen/electron carrier system

The diagram shows the relation between the carriers and the production of ATP.

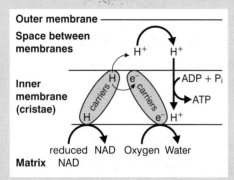

- Hydrogen released during glycolysis, the link reaction and Krebs cycle combines with carriers in the cristae.

- The hydrogen is split into protons and electrons ($H \rightarrow H^+ + e^-$)

- The protons are passed into the space between the two mithochondrial membranes and the electrons returned to the inner surface of the cristae.

- When the protons return through the membrane the energy released is used to produce AIP.

- As the protons move through the membrane they combine with the electrons and oxygen to form water.

Using different respiratory substrates

- The organic substance that is the **starting point** for respiration is called a **respiratory substrate**. It can be a lipid, protein or carbohydrate.

- One way to find which substrate is being used it to calculate **respiratory quotient** (RQ):

$$RQ = \frac{\text{amount of } CO_2 \text{ produced}}{\text{amount of oxygen consumed}}$$

- Example calculation for respiration of a fat:

$$2C_{51}H_{98}O_6 + 145O_2 \rightarrow 102CO_2 + 98H_2O$$

$$RQ = 102/145 \qquad = 0.7$$

- RQs for some important substrates:

lipids/fats:	0.7
protein	0.9
carbohydrates	1.0

- If RQ is greater than 1.0 anaerobic respiration must be involved.

Anaerobic respiration

In some circumstances there is not enough oxygen for the cell to respire aerobically. To produce ATP under these conditions the cell must respire anaerobically.

- Anaerobic respiration relies on glycolysis only to produce ATP, so it is far less efficient than aerobic respiration.

- A single molecule of glucose can produce up to 38 molecules of ATP in aerobic respiration – it produces only 2 in anaerobic respiration.

- Reduced NAD is normally reconverted to NAD when the hydrogen is passed into the mitochondria. If the cell is unable to respire aerobically this cannot happen.

- If all the NAD were reduced, glycolysis could not continue.

- In anaerobic respiration pyruvate removes the hydrogen from the reduced NAD, allowing NAD to be reformed.

- In doing so the pyruvate is converted to lactate (animals) or ethanol and CO_2 (plants).

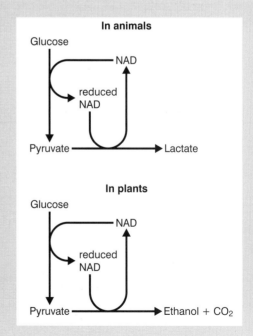

The Sheffield
College

Hillsborough LRC
Telephone: 0114 260 2254

Examiner's hints

Use the information from the stem of the question to recall what you know.

Mitochondria are being used, so ...

- this is where aerobic respiration takes place
- Krebs cycle and the hydrogen carrier system are involved
- organic molecules are broken down to carbon dioxide and hydrogen
- NAD and FAD take up the hydrogen and are reduced
- ATP is made from ADP and phosphate
- oxygen is the terminal acceptor of hydrogen
- water is formed
- Summary:

Going into the mitochondria	Coming out of the mitochondria
pyruvate	carbon dioxide
ADP and phosphate	ATP
oxygen	water

Q1

A preparation of mitochondria was made from liver tissue. Substances were added to this preparation and the amount of oxygen remaining in the preparation was monitored over a period of time. The diagram shows the trace obtained and the times when the different substances were added.

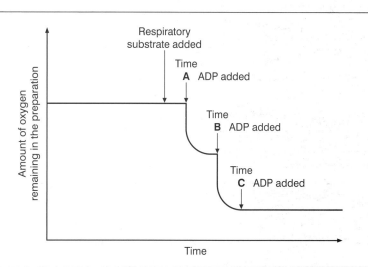

(a) Suggest why the respiratory substrate added to this preparation was a molecule from Krebs cycle and not glucose.

.... [1 mark]

(b) What additional substance, other than those mentioned on the diagram, would need to be added to this preparation in order to get the results shown?

.... [1 mark]

(c) Explain:

(i) why the amount of oxygen fell between times **A** and **B**;

....

[2 marks]

(ii) the shape of the trace after time C.

....

.... [1 mark]

[Total 5 marks]

Examiner's hints

● This question refers to the stages of respiration but the hydrogen carrier system seems to have been left out. Oxidative phosphorylation is a process within this stage and has been used here instead.
● To simplify things anaerobic respiration and glycolysis have also been separated but remember glycolysis occurs at the beginning of both aerobic and anaerobic respiration.
● Have a summary of each of the stages clear in your mind.

	Starts with	Ends with	What is made				
			ATP	reduced NAD	NAD	reduced FAD	CO_2
Aerobic respiration							
Glycolysis	Glucose	Pyruvate	✓	✓			
Link reaction	Pyruvate	Acetyl coenzyme A		✓			✓
Krebs	Acetyl coenzyme A	4 carbon molecule – oxaloacetate	✓	✓		✓	✓
Hydrogen carrier system (oxidative phosphorylation)	Reduced NAD or reduced FAD	Water	✓				
Anaerobic respiration							
Glycolysis	Glucose	Pyruvate	✓	✓			
Fermentation in animals	Pyruvate	Lactate			✓		
Fermentation in plants	Pyruvate	Ethanol			✓		✓

Q2

The diagram shows the main steps in the biochemical pathways involved in respiration.

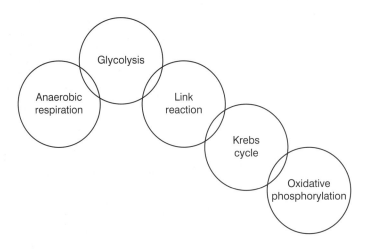

(a) (i) Which step or steps take place on the cristae of the mitochondria?

....

[1 mark]

(ii) In which step or steps is carbon dioxide produced in an animal cell?

....

[1 mark]

(b) If a pond freezes over during the winter, goldfish can remain alive in the water under the ice. Explain why they use the carbohydrate stores in their bodies much faster in these conditions.

....

....

[2 marks]

(c) What is the main difference between the way in which ATP is produced by oxidative phosphorylation and the way in which it is produced in photosynthesis?

....

....

[1 mark]

[Total 5 marks]

Answers to Questions to try are on pages 87–88.

5 Water transport in plants

Exam question and student's answer

Answers should be written in continuous prose. Credit will be given for biological accuracy, the organisation and presentation of the information and the way in which the answer is expressed.

The diagram shows the main pathways by which water moves through a plant.

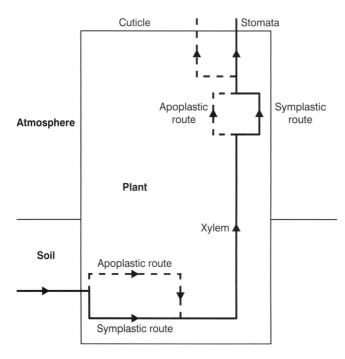

(a) Explain how water moves from the soil into the xylem of the root.

> Roots have special epidermal cells with root hairs. These hairs increase the surface ✓ area of the root and so more water can get in. Water moves into the walls of the cells of the root this is the apoplastic ✓ route. It can also get into the cytoplasm this is called the symplastic ✓ route. In both cases this movement is from an area of lower water potential to an area of higher water ✓ potential. Water will move toward the xylem where it is being removed. As there is less water at the centre of the root where xylem is found a concentration gradient will exist from the outside to the inside along which the water will move.

4/6

[6 marks]

34

(b) Explain how water moves through the xylem in the stem of the plant.

Water evaporates from the leaf this is called transpiration. As it ✓ does this it pulls water out of the xylem to take the place of the water leaving the leaf. Each molecule of water is sticky and sticks to the ✓ molecule next to it. This is known as cohesion of water. So pulling water from the xylem will pull up the water underneath it and a whole column ✓ will be moved up the xylem. This is like pulling at the end of a pencil in a tube. As you pull the top of the pencil the rest of the pencil will move too.

[5 marks]   ③/5

(c) Explain how the structure of a leaf allows efficient gas exchange but also limits water loss.

Gas exchange takes place through the stomata. Stomata are holes in the leaf which allow gases like oxygen and carbon dioxide to enter or leave. Carbon dioxide is used for photosynthesis when it combines with water which comes from the roots. Light and chlorophyll are also necessary for photosynthesis to occur. The problem is that this process will only take place during the daylight and as carbon dioxide gets into the leaf, water will leave. To stop too much water leaving the leaf the stomata can ✓ be closed. This will normally take place when there is not enough water inside the leaf or when a lot of water is ✓ being lost. The outside of the leaf is also covered by a ✓ waxy layer called the cuticle ✓ which will not let water through.

[6 marks]   ④/6

[Total 17 marks]   11/17

Questions to try

Examiner's hints
- Only water movement is included in all syllabuses. The only way that the rate of water movement can be measured is by using a potometer. The values you get will only allow you to compare a number of different conditions on one plant.
- You will be expected to have used a potometer and to know how to set one up.
- Remember the conditions that could affect the rate of transpiration – air speed, air humidity, temperature. Be prepared to explain graphs showing rates of transpiration as those factors are changed.

Q1

The diagram represents one type of potometer often used in transpiration studies.

Graduated capillary tube

(a) **(i)** What is understood by the term *transpiration*?

....

....

....

[1 mark]

(ii) When using this apparatus to measure the rate of transpiration, what major assumption must be made?

....

.... [1 mark]

(b) What measurements must be made in order to determine the rate of water movement?

....

.... [1 mark]

(c) Why is it advisable to cut the end of the shoot under water before it is placed in the apparatus?

....

.... [1 mark]

(d) Suggest **one** method by which the rate at which water rises up the xylem of a tall forest tree might be measured.

....

.... [1 mark]

[Total 5 marks]

Examiner's hints
● You will be expected to look at unfamiliar information and draw conclusions from the principles you know. Think about the effect of blowing water into and sucking it out of the xylem cells.
● Capillarity and root pressure will move water up to a metre and so the positive pressure (blowing water into the cells) involved is very low.
● Transpiration will move water many metres and so the negative pressure (sucking water out of the cells) involved is very high.

Q2

The diameter of a branch from a small tree was measured over a 24-hour period. The results are shown in **Graph 1**.

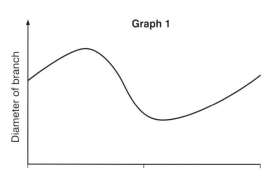

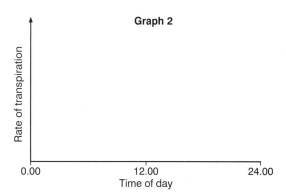

(a) Sketch a curve on the pair of axes in **Graph 2** to show the rate of transpiration for the same 24-hour period as **Graph 1**.

[1 mark]

(b) The following have been used to explain the movement of water in xylem:

A cohesion/tension;

B root pressure;

C capillarity.

(i) Which of these is best supported by the evidence in **Graph 1**?

....

[1 mark]

(ii) Explain your answer.

....

....

....

[2 marks]

[Total 4 marks]

Answers to Questions to try are on page 89.

Exam question and student's answer

The diagram shows how the concentration of glucose in the blood is regulated.

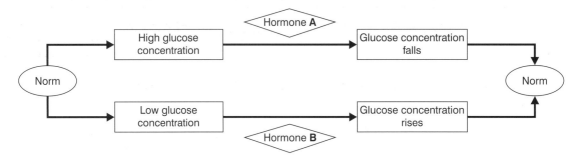

(a) Name

(i) hormone **A**, ___Insulin ✓___ 1/1

[1 mark]

(ii) hormone **B**. ___Glycogon___ 0/1

[1 mark]

(b) Explain how hormone B brings about the change shown in the diagram.

___Glygogen in the liver is changed to glucose.___ ✓ 1/2

[2 marks]

(c) Two people fasted for 12 hours and each was then given a drink which contained 100 g of glucose. Their blood concentration was measured regularly for 3 hours. The results of the investigation are shown in the table.

Time after glucose drink/minutes	Blood glucose concentration/mg per 100 cm³ of blood	
	Person **X**	Person **Y**
0	81	90
20	136	131
40	181	142
60	213	89
90	204	79
120	147	74
150	129	86
180	113	89

(i) Suggest an explanation for the changes in the blood sugar level of person **X**.

___Person X is diabetic. ∧___ 0/1

[1 mark]

(ii) Explain how the concentration of hormone A in the blood would vary in person **Y** between 0 and 60 minutes.

High blood glucose levels are detected in the pancreas which ✓ secretes insulin. So the level of insulin increases. ✓

2/2

[2 marks]

(iii) Suggest an explanation for the results shown by person **Y** between 90 and 180 minutes.

Blood glucose level falls too low, glycogon is released which causes ✓ it to increase.

1/1

[1 mark]

[Total 8 marks] 5/8

How to score full marks

🎯 Spelling counts

In **(a)(ii)** and in **(b)** this student produced new words that are mixtures of glycogen and glucagon.

Glucose — a **monosaccharide** – the soluble carbohydrate carried in the blood.

Glycogen — a **polysaccharide** – the insoluble carbohydrate (made of many glucose molecules joined together) stored in the liver and the muscles.

Glucagon — a **hormone** – produced in the pancreas and increases the blood glucose level.

🎯 Explain the reason

In **(c)(i)** the student thought he had done enough to get the mark, and logically the fact that person X is diabetic is the reason why the blood glucose level increases. This does not, however, **explain** the changes. Failure to control blood glucose levels is due to the fact that a diabetic produces small amounts of insulin (or none) – the expected answer.

🎯 Give a reasoned argument

Part **(c)(ii)** asks not only **how** the concentration of insulin changes but also **why**. The student correctly explained that high blood glucose levels would induce the secretion of insulin and then linked the increased concentration of that hormone with the observed change of blood glucose level. This was a good answer.

🎯 Linked ideas

Sometimes, rather than a fact **a link is expected**. In part **(c)(iii)** the link was between the release of glucagon (the student has kept to the same spelling – glycogon – so we can presume he means glucagon) and the conversion of glycogen to glucose. The effect of this is to raise the blood glucose level (as we can see from the table). The student almost got the link but did not suggest where the extra glucose came from.

Don't make these mistakes...

Don't misspell technical terms and chemicals. Although your work will be marked for content and any word you use will be read phonetically, there are places where the correct spelling is vital – don't mix up words like glucose, glycogen and glucagon.

Don't confuse describe and explain. Describe means 'what is the change?'. Explain means 'what is the cause of the change?'

Be consistent. If you are unsure of the spelling of a term or chemical commit yourself to one and stick to it all the way through. The examiner will be clear on what the word is intended to be and might not penalise you if you have to use this term in further sections of the question.

Key points to remember

Homeostasis involves keeping conditions inside an organism constant.

Why?

- Reactions in the body are controlled by enzymes. Changes in pH and temperature affect the rate of these enzyme-controlled reactions. Outside the optimum range enzymes won't function properly.

- External temperature changes a great deal. A constant internal environment allows independence from these changes. Animals can thus live in conditions as cold as the Arctic and as hot as the tropics.

- Water moves in and out of the cells by osmosis. By maintaining a constant water potential in the fluid surrounding their cells, osmotic problems are avoided.

Negative feedback

- Variables such as temperature, pH and the concentration of many dissolved substances have a set or normal level (often known as the norm).

- Negative feedback is **the reaction to a change from this set level, which will return the variable to its norm value**. This can be summarised as:

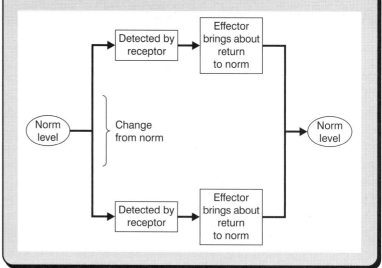

Control of body temperature

- Variables can change in either direction – the temperature of the blood can increase or decrease.

- To return the temperature to the norm must involve processes that can increase or decrease the temperature.

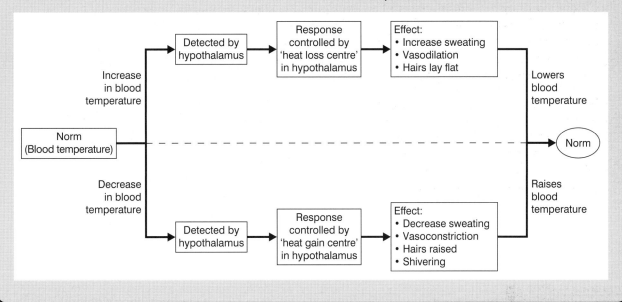

Control of blood glucose concentration

Insulin

- A hormone produced by the β cells of the islets of Langerhans in the pancreas.

- Secretion is stimulated by the rise in blood glucose.

- Speeds up the rate at which glucose is taken into liver and muscle cells from the blood.

- Activates an enzyme which is responsible for conversion of glucose into glycogen.

Glucagon

- A hormone produced by the α cells of the islets of Langerhans in the pancreas.

- Secretion is stimulated by the fall in blood glucose.

- Activates an enzyme that is responsible for the conversion of glycogen into glucose.

- Stimulates the formation of glucose from other molecules such as amino acids.

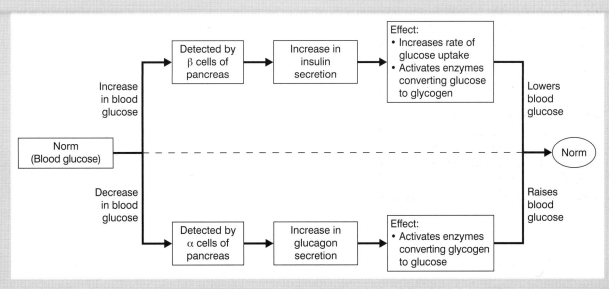

Questions to try

Q1

The diagram summarises the mechanism of temperature regulation in a mammal.

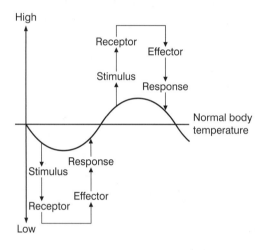

(a) Use the information in this diagram to explain what is meant by *negative feedback*.

....

....

....

....

[1 mark]

(b) Where are the receptors which detect a rise in the temperature of the blood?

....

[1 mark]

(c) Describe the part played by the blood vessels in the skin in returning a high body temperature to its normal value.

....

....

....

[2 marks]

(d) Give one advantage to the body of possessing separate mechanisms to return high and low body temperatures to their normal value.

....

[1 mark]

[Total 5 marks]

Q2

The extract below is an entry from a dictionary of biological terms.

> **blood glucose pool**: the total amount of glucose in the blood at any one time. Accurate control of blood glucose is very important. If the concentration falls too low, the central nervous system ceases to function correctly. If it rises too high, then there will be a loss of glucose from the body in the urine. Hormones including *insulin* and *glucagon* play an important part in maintaining a constant level of glucose in the blood. Although the overall levels stays within narrow limits, it is important to realise that glucose is always being added to and removed from the blood glucose pool. Digestion and absorption of carbohydrates and conversion from the body's stores of glycogen and fats tend to increase the blood glucose level while such processes as respiration decrease it. This is summarised in the diagram.

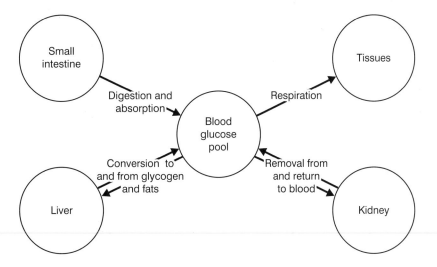

(a) Describe the part played by insulin and glucagon in maintaining a constant level of glucose in the blood.

[6 marks]

(b) In the kidney, glucose is removed from and returned to the blood glucose pool. Explain how this occurs.

[6 marks]

[Total 12 marks]

Answers to Questions to try are on pages 89–90.

Producing concentrated urine – loop of Henle and collecting duct

1 The **descending limb** of the loop of Henle is permeable to water but not sodium and chloride ions. Water will move out of the nephron by osmosis down a water potential gradient.

2 The loss of water causes the ion concentration to increase as the fluid passes down the descending limb. Concentration is highest at the apex.

3 The **ascending limb** of the loop of Henle is permeable to sodium and chloride ions but impermeable to water. These ions are lost by diffusion but are also pumped out by active transport.

4 Pumping the ions out increases the concentration gradient in the tissue of the medulla surrounding the tubules.

5 The **collecting duct** is permeable to water. Since the concentration of ions is always greater in the surrounding medulla tissue than in the collecting duct, water will move out of the collecting duct by osmosis.

6 The concentration gradient in the tissue of the medulla ensures that there is always a water potential gradient between the collecting duct and the surrounding tissue. Water will continue to be removed and this will result in concentrated urine.

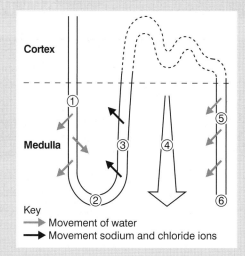

Cortex

Medulla

Key
→ Movement of water
➜ Movement sodium and chloride ions

The kidney and water balance

● In very hot conditions, large amounts of water are lost as sweat. If there is insufficient drinking water to make up for this loss, the water concentration of the blood will fall.

● **Antidiuretic hormone** (ADH) plays an important role in conserving water in such circumstances.

● ADH is secreted by the **pituitary gland** and acts by increasing the permeability to water of the **second convoluted tubule** and the **collecting duct**.

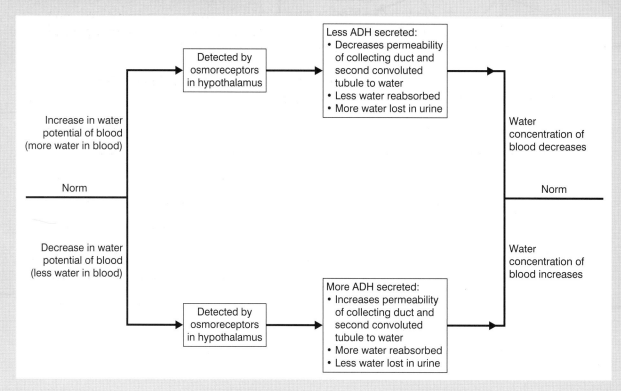

Increase in water potential of blood (more water in blood) → Detected by osmoreceptors in hypothalamus → Less ADH secreted:
• Decreases permeability of collecting duct and second convoluted tubule to water
• Less water reabsorbed
• More water lost in urine
→ Water concentration of blood decreases

Norm — Norm

Decrease in water potential of blood (less water in blood) → Detected by osmoreceptors in hypothalamus → More ADH secreted:
• Increases permeability of collecting duct and second convoluted tubule to water
• More water reabsorbed
• Less water lost in urine
→ Water concentration of blood increases

Questions to try

Q1

Answers should be written in continuous prose. Credit will be given for biological accuracy, the organisation and presentation of the information and the way in which the answer is expressed.

The diagram shows the main functions of the different parts of a kidney tubule.

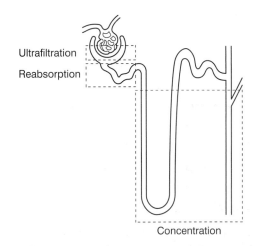

Kidney failure may result in a build up of urea in the blood, protein in the urine and retention of tissue fluid. Patients with kidney failure may need a transplant. Recipients of transplanted kidneys are treated with drugs that suppress their immune system.

(a) Explain how ultrafiltration and reabsorption remove urea from the blood without losing essential nutrients such as protein and glucose.

....

....

....

....

....

....

....

....

....

....

[5 marks]

(b) Explain the part played by the loop of Henle and the collecting duct in concentrating urine in a healthy individual.

....

....

....

....

....

....

....

....

....

[6 marks]

(c) Explain the link between protein in the urine and retention of tissue fluid in patients with kidney failure.

....

....

....

....

....

[3 marks]

(d) If not treated with drugs, explain how the recipient's immune system would reject a transplanted kidney.

....

....

....

....

....

....

[3 marks]

[Total 17 marks]

Examiner's hints

Remember what happens in each of the regions of the nephron:
- renal capsule – ultrafiltration
- first convoluted tubule – reabsorption of glucose and amino acids
- loop of Henle – creation of a concentration gradient down the medulla
- second convoluted tubule – reabsorption of water
- collecting duct – reabsorption of water.

Q2

The diagram shows the structure of a nephron (kidney tubule).

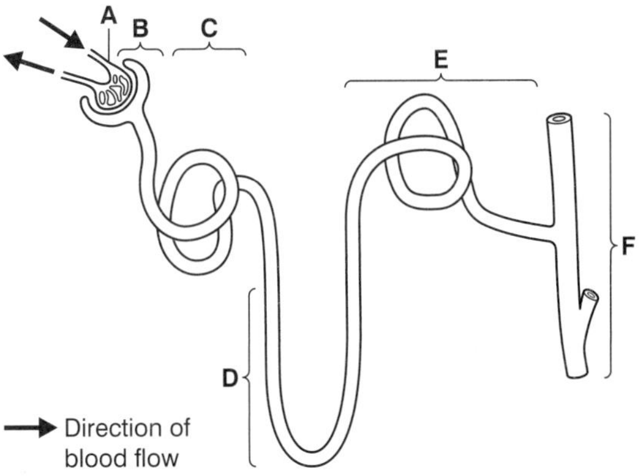

(a) Name the major artery of which **A** is a branch.

....

[1 mark]

Direction of blood flow

(b) Name the process that takes place in the part of the nephron labelled **B**.

....

[1 mark]

(c) Give the letter or letters which represent the region or regions of glucose reabsorption.

....

[1 mark]

(d) If humans drank only sea water, they would be unable to remove the excess salt from their bodies. Marine mammals, such as seals and whales, are able to remove excess salt because region **D** is relatively longer in these mammals than in humans. Explain why the longer length of region **D** enables marine mammals to remove all of the excess salt from their bodies.

....

....

....

....

....

....

[3 marks]

[Total 6 marks]

Answers to Questions to try are on pages 90–92.

8 Digestion

The diagram represents a small part of the duodenum wall seen in section using a microscope.

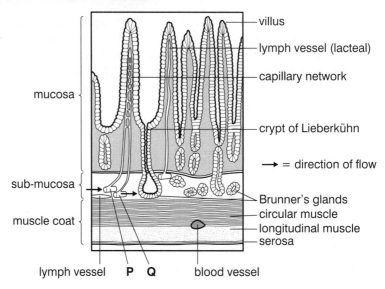

mucosa

villus

lymph vessel (lacteal)

capillary network

crypt of Lieberkühn

→ = direction of flow

sub-mucosa

muscle coat

Brunner's glands
circular muscle
longitudinal muscle
serosa

lymph vessel **P Q** blood vessel

(a) (i) How is the epithelium (inner surface) of the duodenum adapted for absorption of soluble products of digestion?

> It is thrown into a number of finger-like projections called villi which increases the surface area.

[2 marks] ¹/₂

(ii) Suggest two other functions of the epithelium layer.

> Secretion of mucus. Absorption of the lipids.

[2 marks} ¹/₂

(b) What is the importance of the muscle in the duodenum wall?

> Moves food along the duodenum.

[1 mark] ¹/₁

(c) Briefly give **four** ways in which the blood composition at site P differs from the blood composition at Q.

> P has less glucose, less amino acids, less lipid, less oxygen, more carbon dioxide.

[2 marks] ¹/₂

(d) Along which vessel will blood at Q pass on leaving the duodenum?

> Hepatic portal vein ✓

[1 mark] ¹/₁

[Total 8 marks] ⁵/₈

How to score full marks

Know the difference between similar terms

Villi and microvilli are very different! You will **never get marks** for using the **wrong technical terms** – no matter how similar the words. **Villi** are finger-like projections of the **inner layer of the intestine**, the mucosa, but **microvilli** are projections of the **cell membrane**. In part **(a) (i)** the student wrote villi but should have written microvilli.

Remember the principles

The rate of diffusion (and therefore facilitated diffusion) is affected by four things:

1 The **surface area** must be '**large**'.
2 The **diffusion pathway** must be '**short**'.
3 There must be some mechanism to ensure that a **concentration gradient** is '**maintained**'.
4 Increased temperature will increase the kinetic energy of the molecules involved so '**more collisions**' with the membrane will occur, which increases the possibilities of molecules being absorbed.

Look carefully at the question and identify which of these things refer to the surface. In part **(a) (i)** only the first two are relevant.

Read the question

In part **(a) (ii)** the question asks clearly for two 'other' functions but the student gave absorption as one of the alternatives – which is the function given in part **(a) (i)**. He did try to name a molecule that is not soluble, lipid, but missed the point that the process is the same. A silly mistake!

Only give the number of facts you are asked for

The student has given five ways, instead of the four asked for in the question. One of these is wrong, so he recieved only one mark, instead of the possible two. **Never give more facts than you are asked for.**

Expect links

One of the differences between AS and A2 is that you will be expected to be able to **bring together different areas** of the subject. So when you are revising **look for links**. In this topic – digestion – you can expect to be asked about the structure of organic molecules, about enzymes and about blood vessels that carry blood to and away from the gut. Part **(d)** is simply recall of these facts.

Don't make these mistakes...

Don't give more facts than asked for. Everything you write will be considered but the examiner will not select the correct answers from your list – and you could be penalised if you make any mistakes.

When answering a question make sure you **focus on the correct side of the argument.**

Don't get terms that are similar muddled up – using the wrong term won't get you any marks!

Key points to remember

What is digestion?

- Digestion is the breakdown of large, insoluble molecules into small, soluble molecules that can be absorbed by the body.

- Digestion is one part of a series of processes by which organisms obtain organic molecules.

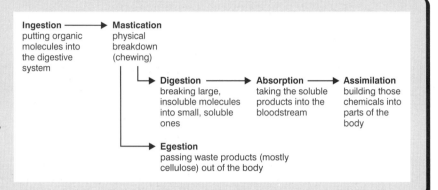

Ingestion → **Mastication**
putting organic molecules into the digestive system
physical breakdown (chewing)

Digestion → **Absorption** → **Assimilation**
breaking large, insoluble molecules into small, soluble ones
taking the soluble products into the bloodstream
building those chemicals into parts of the body

Egestion
passing waste products (mostly cellulose) out of the body

Enzymes

- To make digestion occur quickly without increasing the temperature too much and to ensure that they occur in the right order, enzymes control all the reactions.

- All digestive enzymes are hydrolytic enzymes. They break down large molecules into smaller ones by the addition of water.

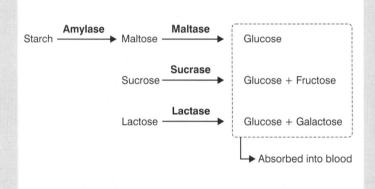

Starch — **Amylase** → Maltose — **Maltase** → Glucose

Sucrose — **Sucrase** → Glucose + Fructose

Lactose — **Lactase** → Glucose + Galactose

→ Absorbed into blood

Amylase — Buccal cavity

Oesophagus

Pepsin — Stomach

Trypsin Amylase Lipase — Pancreas

Dipeptidases Disaccharidases — Small intestine

Colon

Rectum

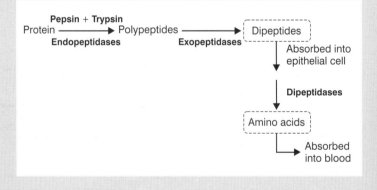

Pepsin + Trypsin
Protein ——→ Polypeptides ——→ Dipeptides
Endopeptidases **Exopeptidases**
Absorbed into epithelial cell

↓

Dipeptidases

↓

Amino acids
Absorbed into blood

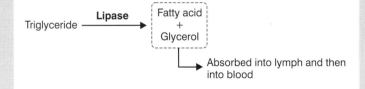

Triglyceride — **Lipase** → Fatty acid + Glycerol
Absorbed into lymph and then into blood

Features of the ileum

The ileum is the area of the gut involved in absorption.

Feature of ileum	Reason for feature
Long tube	Large surface area for absorption
Folded wall	Large surface area for absorption
Inner layer has finger like projections – Villi	Large surface area for absorption
Epithelium has microvilli	Large surface area for absorption
Epithelium has many mitochondria	Provides energy for active transport
Villi contain capillaries	Remove absorbed products of carbohydrate and protein digestion
Villi contain lacteals	Remove absorbed products of lipid digestion

Control of secretions

Stimulus	Nervous/hormonal	Effect	Area where secretion acts
Sight/smell/thought of food	Nervous – conditioned reflex	Secretion of saliva	Buccal cavity
Food in the mouth	Nervous – reflex	Secretion of saliva	Buccal cavity
Food in the mouth	Nervous – reflex	Secretion of gastric juice	Stomach
Food in stomach – distending stomach	Nervous – reflex	Secretion of gastric juice	Stomach
Food in stomach	Hormonal – *gastrin*	Secretion of gastric juice	Stomach
Food in small intestine	Hormonal – *secretin*	Secretion of pancreatic juice – containing alkaline fluid Synthesis of bile by liver	Small intestine
Food in small intestine	Hormonal – *pancreozymin*	Secretion of pancreatic juice – containing enzymes	Small intestine
Food in small intestine	Hormonal – *cholecystokinin*	Release of bile from gall bladder	Small intestine

Questions to try

Q1

The drawings show the surface of epithelial cells lining the small intestine of a rabbit. The first drawing is from a healthy animal and the second is from an animal whose intestine has been invaded by pathogenic *Escherichia coli* bacteria.

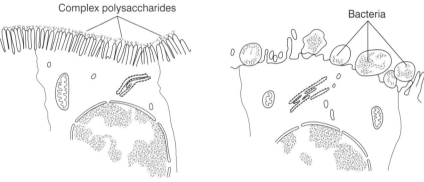

Healthy animal   Animal invaded by *E. coli*

(a) Describe the effect the *E. coli* bacteria had on the surface of the epithelial cells.

....

[1 mark]

(b) Describe and explain the effect the *E. coli* would have on:

  (i) digestion of dipeptides in the small intestine;

....

....

  (ii) absorption of the products of this digestion.

....

....

[3 marks]

(c) It has been suggested that invasion of *E. coli* would make further damage to the epithelial cells more likely due to the activity of protein-digesting and lipid-digesting enzymes released by the pancreas. Using information from the drawings, suggest why this might occur.

....

....

[2 marks]

[Total 6 marks]

Q2

(a) The diagram shows a section through part of the wall of the small intestine.

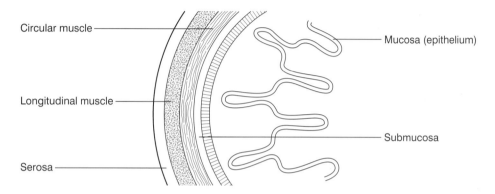

Describe how the muscle layers cause food to move along the small intestine.

....

....

....

....

[2 marks]

(b) The secretion of digestive juices into the mammalian gut is controlled by the endocrine and nervous systems. Complete the table about the secretion of digestive juices.

Stimulus that triggers secretion	Effect	Digestive juice secreted
....	Parasympathetic nerve stimulates salivary gland	Saliva
Contact of food with stomach lining	.... secreted	Gastric juice
Contact of food with duodenum lining	Cholecystokinin secreted	....
Contact of food with duodenum lining	.... secreted	Alkaline fluid from pancreas

[4 marks]

[Total 6 marks]

Answers to Questions to try are on pages 92–93.

Exam question and student's answer

The graph shows the changes in permeability of an axon membrane to sodium and potassium ions during an action potential in a neurone.

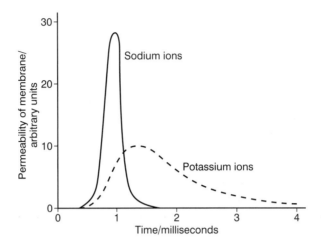

(a) Use information in the graph to explain why, at the start of an action potential, the potential difference across the membrane rapidly changes from negative to positive.

The permeability of the membrane increases ✓ and sodium rushes into the cell.

[2 marks]

¹/₂

(b) Suggest why, during a period of intense nervous activity, the metabolic rate of a nerve cell increases.

If the nerve is working harder it will <u>need</u> more energy.

[2 marks]

0/₂

(c) Predict the effect on an action potential of lowering the external concentration of sodium ions. Explain your answer.

Less sodium enters ✓ the cell and the action potential may be lower. ✓

[2 marks]

²/₂

[Total 6 marks]

³/₆

How to score full marks

What does the graph tell you?

You have learnt that an **action potential** is caused by **sodium ions** moving **into** the neurone and **potassium ions** moving **out**. These ionic movements take place when sodium or potassium **gates** in the membrane open, **increasing the permeability** to those ions. In part (a) the student correctly related increasing permeability to the direction of sodium ion movement. However, as he referred to sodium, not sodium ions, no credit was given.

'Suggest' – rewrite the question yourself

In part **(b)** increased metabolic rate implies greater ATP production so the question really asks 'why is more ATP needed in an active neurone?' The student's answer is too vague; **you must be logical and use A level quality material**. As the cation pump (sodium/potassium pump) is moving ions against their concentration gradient, this has to be 'active transport', which requires ATP. This pump is vital in re-establishing a resting potential following an action potential. The **more active neurones** will be generating **more action potentials** and therefore the **cation pump will be more active**, needing **more ATP**.

For 'predict' read 'make an educated guess'

You may never have met this situation before – it may even be a hypothetical one – but **use your understanding** of the process **to guess an answer**. The student does that very effectively in part **(c)**. As sodium ions move by diffusion and the rate of diffusion is affected by the difference in concentration (Fick's Law) you could also deduce that the rate of movement of ions would be slower – so perhaps the slope of the action potential would be shallower, so that it would take longer to reach the action potential.

Don't make these mistakes...

Always be accurate with chemical terms. If you write 'sodium' it means an atom of the element sodium. In a neurone, if an atom moved there would be no change in potential difference (charge) across the membrane, so the fact that sodium ions move is vital.

Cells, tissues or organs do not 'want' or 'think' or 'feel'. Avoid using words like these in your answers.

Key points to remember

Important terms to understand

- **Stimulus** – signal to which an organism responds. Can be **external** (e.g. sound) or **internal** (e.g. blood temperature).

- **Receptor** – a cell or organ that detects a stimulus. **Specific** – will only respond to one type of stimulus. **Sensitive** – able to detect very small changes. They convert a stimulus into a nerve impulse.

- **Effector** – an organ that responds to an impulse from the nervous system.

- **Response** – activity in a muscle or gland that is the result of a nervous impulse.

Transmission of nervous impulses

Changes in electrical potential across the plasma membrane pass along a neurone.

1 Resting potential (neurone at rest)

- **Higher concentration** of potassium ions (**K⁺**) and **lower concentration** of sodium ions (**Na⁺**) in the cytoplasm.
- K⁺ diffuse out easily.
- Na⁺ diffuse in more slowly.
- So there are fewer positive ions inside the membrane than outside.

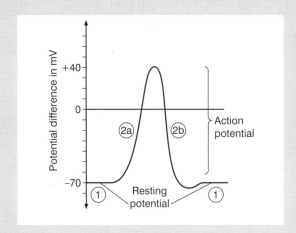

Potential difference across the membrane is −70mV.

2 Action potential

(a) *Depolarisation (impulse is started)*

- Sodium channels/gates open.
- Na⁺ enter the neurone.
- So there are more positive ions inside the neurone than outside.
- **Potential difference across the cell membrane is +40 mV**.
- Sodium channels/gates close.

(b) *Repolarisation (returning to rest)*

- Potassium channels/gates open.
- K⁺ leave the neurone.
- So there are fewer positive ions inside the neurone than outside.
- **Potential difference cell membrane is −70 mV.**

An active transport mechanism (the **cation** or **sodium/potassium pump**) in the plasma membrane pumps out the Na⁺ that entered the cell during the action potential and pump K⁺ in.

Threshold

- The threshold value is the **lowest level of stimulus** that will trigger an action potential.
- When a nerve cell is stimulated the plasma membrane must be **depolarised enough** to open some of the sodium gates.
- If the stimulus is **too small** this will not happen and **an action potential will not take place**.

All or nothing

- Once the threshold is overcome potential difference changes.
- The action potential is always the **same size** in the same neurone
- Either an action potential will be generated or not – 'all or nothing'.
- You are aware of the different **strengths** of stimulus – e.g. louder or softer noises – only because the **frequency** of impulses changes.
- In any neurone all action potentials are the same strength and all action potentials travel at the same speed.

A nerve impulse results from the movement of an action potential along a neurone

- The change of potential difference (action potential) at one point causes some Na^+ gates in the next section of the neurone to open.
- The membrane of the next section of neurone becomes depolarised.
- The threshold is overcome.
- An action potential occurs there.
- The membrane of the next section of the neurone becomes depolarised.
- And so on ...

Saltatory conduction

- Some neurones are surrounded by a myelin sheath.
- Depolarisation can only occur where there are gaps in the sheath.
- The action potential jumps from one gap to the next.
- Speed of the impulse transmission (up to $120 ms^{-1}$) is many times greater than in unmyelinated neurones (as low as $0.5 m s^{-1}$).

Synapse

- The nervous system is made of many cells that communicate without physically joining.
- The space between two neurones is a synapse.
- Chemicals (neurotransmitters) cross the synapse and depolarise a neurone, starting an action potential.
- More than two neurones may meet at a synapse.

Functions of the synapse:

- **'One-way valve'** – the chemical is produced on only one side of the synapse, so the impulse always goes the same way down the same neurone.
- **Threshold** – enough chemical has to be produced at the synapse before a connection will be made.
- **Summation** – weak stimuli from many neurones can have an additive effect, producing enough chemical to make the link.
- **Inhibition** – some neurones that make up the synapse can produce a chemical that inhibits depolarisation.

Synaptic transmission

1 Nerve impulses arrive at presynaptic membrane.

2 Calcium channel proteins in the presynaptic membrane open and allow calcium ions (Ca^{2+}) to diffuse into the neurone.

3 This increases the Ca^{2+} concentration in the synaptic bulb, causing some vesicles to move to and fuse with the presynaptic membrane.

4 The vesicles burst and release the neurotransmitter (acetylcholine) into the synaptic cleft, which it diffuses across.

5 Acetylcholine binds to receptors on the postsynaptic membrane.

6 Sodium channel proteins open, allowing Na^+ to enter the postsynaptic neurone and produce an action potential in that neurone.

7 The acetylcholine is almost immediately hydrolysed into its components by the enzyme acetylcholinesterase. These diffuse back across the synaptic cleft and are absorbed through the presynaptic membrane, and are used to produce more acetylcholine, which is packed into vesicles. Mitochondria in the presynaptic cell provide ATP for resynthesis of the transmitter.

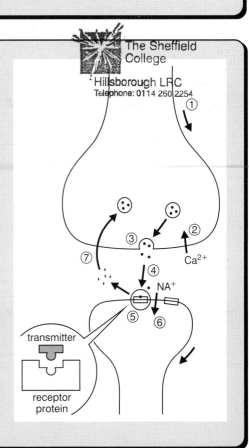

Questions to try

A neurone was suspended in a suitable solution and connected both to an electrical stimulator and to an oscilloscope. The intensity of the stimulus could be varied. The oscilloscope produced a visual record of the action potentials in the neurone.

The diagrams show the apparatus and a summary of the results of the experiment.

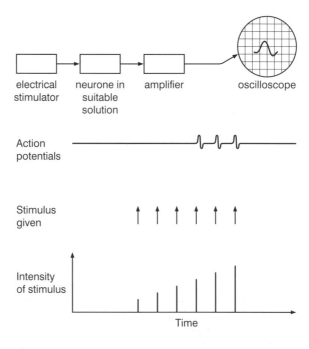

(a) **(i)** What sort of solution would be suitable to use in this experiment?

....

[1 mark]

(ii) Explain why this solution is used.

....

[1 mark]

(b) Explain why the first three stimuli do not produce action potentials.

....

....

....

[1 mark]

(c) **(i)** Give **two** similarities between the action potentials

1 ....

2 ....

[2 marks]

(ii) Sense organs receive stimuli at different intensities. Explain how the neurones transmit this information.

....

....

[1mark]

(d) Explain what happens at a point on a neurone when an action potential is generated and a resting potential is re-established.

....

....

....

....

....

[4 marks]

[Total 10 marks]

Examiner's hints
- The junction between two neurones – a synapse – functions in the same way as the junction between a neurone and a muscle – a nerve–muscle junction.
- It does not matter what neurotransmitter is used in the exam question or what shape is used to represent it. The principle is the same.
- Look at the diagrams carefully and read the keys to see where each chemical is involved in the process. All the information you need will be there: from it you can see how each is different from the normal situation.
- If you are asked to describe and explain, remember you must write what is happening and then why it would happen. In this case it might be easier to explain the effect of the molecule introduced and then describe what this does to the normal functioning of the muscle.

Q2

The diagram on page 68 shows a normal nerve–muscle junction.

Using the evidence from the diagrams, describe and explain the effect on muscle activity of curare, organophosphate and botulin toxin.

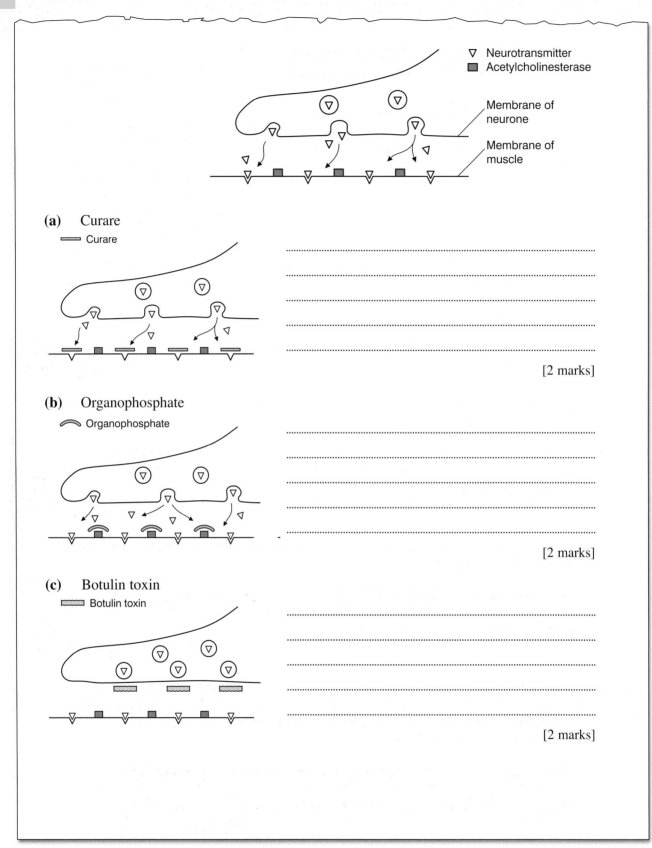

(a) Curare

....

....

....

....

....

[2 marks]

(b) Organophosphate

....

....

....

....

....

[2 marks]

(c) Botulin toxin

....

....

....

....

....

[2 marks]

Answers to Questions to try are on pages 93–94.

10 Transport of respiratory gases

Exam question and student's answer

Fetal haemoglobin has a higher affinity for oxygen than adult haemoglobin. When the partial pressure of oxygen is low, fetal haemoglobin carries much more oxygen. The oxygen-haemoglobin dissociation curves for adult haemoglobin and fetal haemoglobin are shown on the graph.

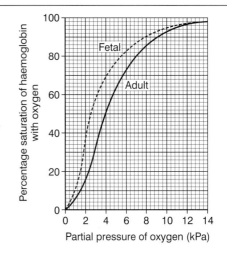

(a) (i) Use the graph to determine the difference in percentage saturation of fetal haemoglobin and adult haemoglobin at 3 kPa.

> Adult 32%
>
> Fetal 60%   Difference 60−32 = 28%   Answer 28 % ✓

 1/1

[1 mark]

(ii) Suggest the advantage of fetal haemoglobin having a higher affinity for oxygen than adult haemoglobin.

> It can take up oxygen when adult haemoglobin is ready to give it up. ∧

 0/1

[1 mark]

(b) As the blood temperature rises the dissociation curve for adult haemoglobin moves to the right. Explain the advantage of this.

> This lowers the affinity ✓ of adult haemoglobin to oxygen, so more ✓ is released to the cells.

 2/2

[2 marks]

(c) Myoglobin is a respiratory pigment similar to haemoglobin. It has a very high affinity for oxygen. Myoglobin is particularly abundant in muscles which produce sustained contractions. Suggest how humans benefit from the presence of myoglobin in muscles used for sustained contractions.

> When muscles are contracting all the time they need lots of energy. This is provided in the form of ATP, produced by respiration. Aerobic respiration produces most ✓ ATP but needs lots of oxygen which the muscle gets from myoglobin.

1/2

[2 marks]

 [Total 6 marks] 4/6

How to score full marks

Show your working

If a numerical answer is needed, usually the correct figure alone will gain you all the marks. If that answer is wrong, examiners will look at your working to see if there is anything to 'salvage'. That is why you will see 'show your working' in some questions. **It is always worth showing how you got to your answer**. In **(a) (i)** the student got the right answer but also proved by his working that he had read the graph correctly.

Write clearly

In part **(a) (ii)** the student used 'it' twice in his answer. Luckily the first 'it' was implied as fetal haemoglobin and the second 'it' as oxygen. **Be sure that you make 'it' clear!**

Use your knowledge

The advantage of fetal haemoglobin having a higher affinity to oxygen is only of importance when the **two types are in contact**. This only occurs at the **placenta**, so a better answer to part **(a) (ii)** would be 'Fetal haemoglobin becomes saturated at lower partial pressures of oxygen which are present in the placenta where the fetus gets its oxygen'.

Use the terms in the question

This question uses lots of **technical terms** – partial pressure, percentage saturation, affinity. **Use them in your answers**, as the student has in part **(b)**.

Simplify the question

When provided with lots of complicated words or long question stems **put the question into your own words and note the clues** that are given. In part **(c)** '… It has a very high affinity for oxygen …' means it carries more oxygen at low partial pressures/will not give up oxygen until there is very little oxygen about. 'Myoglobin is particularly abundant …' (*there is lots of it*) '… in muscles which produce sustained contractions …' (*are active all the time*). **The clue**: active muscles need more oxygen.

Make the links

In **(c)** the student tried to link together a number of ideas, but lost sight of the question, which was '**Why is myoglobin vital in active muscles?**' Marks will go for these links:

● active muscles need oxygen for aerobic respiration

● myoglobin allows this to continue by providing oxygen when the muscle exhausts the available oxygen provided by the blood haemoglobin.

He eventually got the first link but not the second.

Don't make these mistakes...

Do mark lines on a graph to help you **see clearly** what is happening, but if there is more than one graph **label what you have marked**.

Try to avoid the word 'it' in your answers.

Never rewrite the question but always **make the subject of your answer clear** to the examiner.

Know the **meaning**, not just a definition of technical terms.

Partial pressure – the **amount** of oxygen **available**

Percentage saturation – the **amount** of oxygen **being carried**.

Key points to remember

Transport of oxygen

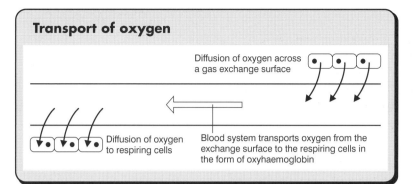

Diffusion of oxygen across a gas exchange surface

Blood system transports oxygen from the exchange surface to the respiring cells in the form of oxyhaemoglobin

Diffusion of oxygen to respiring cells

The oxygen dissociation curve

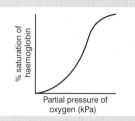

- Percentage saturation of haemoglobin with oxygen is 100% when haemoglobin is carrying all the oxygen it can, in the form of **oxyhaemoglobin**.
- Partial pressure of oxygen means the amount of oxygen present.
- When no oxygen is present the haemoglobin will carry no oxygen. When a certain amount of oxygen is present the haemoglobin will be 100% saturated. Between these two extremes the curve is S-shaped.
- The oxygen dissociation curve can be used to explain how haemoglobin takes up and releases oxygen.

Using the oxygen dissociation curve

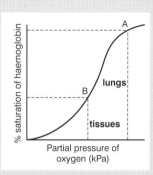

- The **partial pressure** of oxygen is **higher** in the **lungs** than the tissues in the body
- At the partial pressure in the lungs (ignore numbers) haemoglobin is almost completely saturated with oxygen (point A on the curve). The **haemoglobin in blood flowing through the lung capillaries** is **all** converted to **oxyhaemoglobin**.
- In the **tissues** there is **far less oxygen** because it is being **used up in respiration**. The percentage saturation of haemoglobin is lower, so oxygen can be given up to the tissues (point B on the curve.)

Bohr shift

- The body cells require different amounts of oxygen and release different amounts of carbon dioxide depending on their requirements.
- The **ability of haemoglobin to transport oxygen** is affected by the **amount of carbon dioxide present**.
- The **more carbon dioxide**, the more the oxygen dissociation curve is **moved to the right**.
- This is known as the Bohr shift.

In the lungs:
- Partial pressure of **oxygen high**; partial pressure of **carbon dioxide low**.
- **Haemoglobin** will be **fully saturated** with oxygen.

In the tissues:
- Partial pressure of **oxygen low**; partial pressure of **carbon dioxide high**.
- We are now at point C.
- The **percentage saturation** of haemoglobin will be **lower** and **more oxygen will be delivered to the tissues** at the same partial pressure.
- Under these conditions haemoglobin will release oxygen to the tissues at a higher partial pressure: **oxygen will be delivered when there is still oxygen in the tissues**.

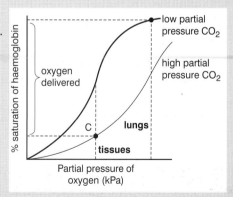

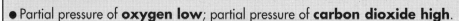

Different dissociation curves

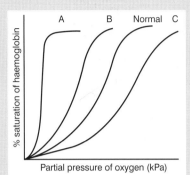

- **Different organisms** have slightly **different** sorts of **haemoglobin**.

- This difference in haemoglobin affects the oxygen-carrying properties of the molecule.

Curve A:

- Pigment with a curve of this shape usually acts as an **oxygen store**.
- The pigment only releases oxygen when the amount of **oxygen in the tissues** is **very low**.
- A respiratory pigment with this curve is **myoglobin**.

Curve B:

- This type of haemoglobin can be **saturated** with oxygen when there is **only a limited amount** available.
- Examples include haemoglobin of a llama and fetal haemoglobin. They both live in environments with relatively low partial pressures of oxygen – llamas live at high altitude; the fetus obtains its oxygen from the placenta, which has a lower partial pressure than the lungs of the mother.

Curve C:

- This haemoglobin **gives up** its oxygen **readily**.
- It is associated with animals that have a **high rate of respiration**. Examples are birds and small mammals such as mice.

Carbon dioxide transport

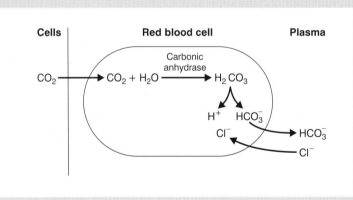

- **Carbon dioxide** diffuses into the red blood cell from the plasma and forms **carbonic acid** (which splits into H^+ and HCO_3^- ions). This reaction is catalysed by the enzyme **carbonic anhydrase**.

- **Hydrogen carbonate ions** (HCO_3^-) diffuse from the cells into the **plasma**.

- **Chloride ions** (Cl^-) diffuse from the plasma into the **red blood cells**. This maintains a neutral charge.

- The **hydrogen ions** (H^+) are taken up by **buffers**. In the red blood cell haemoglobin is the buffer.

- At the lungs the reverse takes place and carbon dioxide is released into the plasma. From here it diffuses into the air spaces of the lung and is exhaled.

Questions to try

The graph shows an oxygen dissociation curve for human haemoglobin.

The loading tension is the partial pressure at which the haemoglobin is 95% saturated with oxygen. The unloading tension is the partial pressure at which the haemoglobin is 50% saturated.

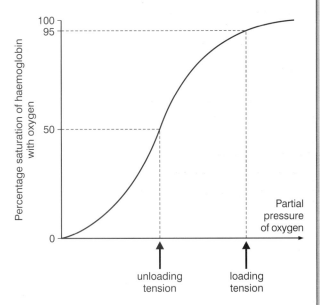

(a) (i) What would be the effect on the unloading tension of an increase in the partial pressure of carbon dioxide?

....

[1 mark]

(ii) Explain how this may be of value in supplying tissues with oxygen.

....

....

[2 marks]

(b) The prairie dog is a small mammal that spends much of its life in an extensive system of burrows where the air may have a low partial pressure of oxygen.

(i) Sketch a curve on the graph which would represent an oxygen dissociation curve for prairie dog haemoglobin.

[1 mark]

(ii) Explain why you have drawn the curve in this position.

....

....

[2 marks]

[Total 6 marks]

Examiner's hints

● If the question asks you to use the information in a graph or table make reference to it in your answer.

● Explain means 'give a reason for' – so you must be able to write why something happens and not simply describe what is happening.

● A dissociation curve that appears to the left means this chemical has a higher affinity for oxygen and is likely to be useful in areas where there is very little oxygen

Q2

A small amount of oxygen diffuses from the blood into the small intestine of a mammal. Some parasitic platyhelminths living in the small intestine can make use of this oxygen.

The graph shows oxygen dissociation curves for human haemoglobin and for the haemoglobin of a parasitic platyhelminth which lives in the human small intestine.

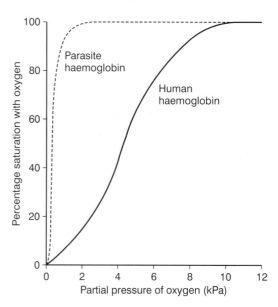

(a) Use the graph to help explain how human haemoglobin releases oxygen when it reaches the cells of the wall of the small intestine.

....

....

....

....

[3 marks]

(b) Explain one advantage to the parasite of having haemoglobin with an oxygen dissociation curve like that shown on the graph.

....

....

....

....

[2 marks]

[Total 5 marks]

Answers to Questions to try are on pages 94–95.

11 Synoptic questions – essay writing

Your A2 exam will include questions which expect you to bring together different content areas of your specification. All exam boards use essays or sections of extended writing for synoptic assessment. The essay is intended to test your ability to bring together material from different areas of the specification and to see if you can organise your ideas and communicate them effectively.

Be prepared

Like any examination technique you need to practise essay writing. There is no point writing an essay with your text book open. So before you begin:

- Review the subject area.
- Read text and lecture notes.
- Make relevant notes.

You need a plan

Write a plan that will act as a skeleton for your essay.

Decide which areas of the specification need to be used:

- Take the title of the essay and briefly write down which areas of the specification you think need to be used.
- Remember it may be material you studied for AS. AS modules of all examining boards cover topics that help you to understand material in the A2 modules. The AS topic 'enzymes', for example, is useful in the A2 topic 'digestion'.

For an essay titled '**The function of proteins**' the following areas would be relevant:

- Diet and digestion in mammals
- Role of haemoglobin
- Principles of immunity and tissue rejection
- Muscle contraction
- The principles of hormonal control
- Properties of cell membranes
- Properties of protein: the importance of long-chain molecules
- Properties of enzyme action

Decide on a structure

Select areas on which to focus your thoughts. This will make sure that the essay covers more than one area of the specification.

You may have been taught to draw spider diagrams to help you brainstorm the major facts to include. These are fine, but they will not make sure you cover **enough areas of the specification**. Try using the idea below:

The types of headings you'll need in your essay can be put into four sets.

Heading set 1	Heading set 2	Heading set 3	Heading set 4
How	Prokaryotes	Cells and biochemistry	Respiration and transport
What	Protoctista	Ecology	Homeostasis and excretion
When	Fungi	Evolution	Nervous system and locomotion
Why	Plants	Plant physiology	Reproduction and growth
Where	Animals	Animal physiology	Nutrition

As you can see, there are five headings per set. Remember these headings. Then, when you come to plan your essay, choose the set of headings that best fit the title of the essay you are considering. You will need to write something for at least three of the headings to cover enough of the specification for a good mark.

For example, for that essay on '**The function of proteins**' the fourth set of headings would fit best with the essay title. Your plan might highlight the following areas:

Respiration and transport	Haemoglobin, antibodies
Homeostasis and excretion	Hormones − insulin
Nervous system and locomotion	Sodium pumps, actin and myosin
Reproduction and growth	Structural proteins
Nutrition	Enzymes, membrane proteins

These topics come from the areas of the specification given on page 00.

If you were asked to write an essay on '**Osmosis and its importance in living organisms**' the first set of headings would probably be the most useful:

How	Diffusion, membrane structure
What	Water movement
When	Translocation, transpiration
Why	Water potential, solute potential
Where	Root, kidney

Remember

- **Do not** write full sentences in your plan.
- **Do not** write a plan as a rough draft of your essay.
- **Do not** spend more than 5 minutes on the plan.
- **Do** brainstorm and write down key words.
- **Do** try to sequence your ideas.

You need to practise

Planning the essay

- Practise writing plans – you need not always write the essay. **Essay plans help clarify facts that may be tested in other questions**.
- The more plans you have thought through – alone or with a friend – the easier it will be to plan the essay you have to write in the examination.
- A good essay is easy to write once you have a good plan.

Writing the essay

- You do not need to write an essay for every plan you prepare when you are revising.
- Always limit writing time to 30–35 minutes.
- Only write the essay when you have revised the topics to be included.
- **Never** write the essay with reference books in front of you. You will **copy** chunks. You will not **understand** the material. You will not **learn** the material.

You need to know how the essay is marked

- The essay is **not** point marked so, unlike all other sections of your exam, you do not have to try to relate 'a fact with a mark'. The essay titles are so broad it would be impossible to write a mark scheme to cover all the facts you could write down.

- The examiner will not expect you to write everything on the topic and maximum marks are based on what examiners expect an 'A' level candidate to do in 30 minutes under the stress of an examination.

- Examiners know you will not remember everything and will possibly make some mistakes but you can still get full marks – we do not expect absolute perfection.

The examiner marks four skill areas:

1 **Scientific content**

The work is marked subjectively – the examiner (and you must get into a habit of doing this too) will place your work into one of three bands:

- **Good** – Your work will contain 'A' level quality factual material and will show that you understand the principles behind it. Any mistakes will be minor.

- **Average** – Most of the work will be of high quality but some areas will be a bit thin and will not really be more than could be expected at GCSE level. There may be some mistakes.

- **Poor** – Most of the work is GCSE level and there are only a few sections that show 'A' level quality work. There are quite a few mistakes.

Even within the same band there will be variations – in the 'good band' some candidates will go into more detail than others or will make fewer mistakes – so in each band the examiner can award one of three marks:

- Good – 16 or 14 or 12.

- Average – 10 or 8 or 6.

- Poor – 4 or 2 or 0.

2 **Breadth of knowledge**

Within this category the examiner considers whether you have brought together relevant material from different areas of the specifications to answer the question.

There are three marks available:

- You will get 3 marks if you have made reference to most of the areas that are mentioned in the specifications related to the essay title.

- You will get 2 marks if you have considered more than one area but have missed out something that is important.

- You will get 1 mark if you have based your work on just one topic.

- You will be given no marks if the material you write is irrelevant.

Primary consumers eat these plants to gain their nutrition. Plants contain all the nutrients a cow would need. However, cows do not have the necessary enzymes to digest and absorb all the nutrients. For example, a cow cannot get its protein directly. Instead it has to have a symbiotic relationship with bacteria in its stomach. The cow itself can digest all the carbohydrates by using amylase and other enzymes. The bacteria is also chemoautotrophic and it breaks down the cellulose and other indigestible plant parts for itself using enzymes. The cow then obtains its nutrients by digesting the bacteria inside its stomach.

Incorrect biology here – a cow *can* get its protein directly, as plant material contains carbohydrate, lipid and protein, but its diet is low in sources of this organic molecule. Thus the bacteria it digests are an important protein source for the herbivore. Although reference is made to bacteria being digested, this point is not clear. No detail is given of the destination of the breakdown products of cellulose or what benefit the bacteria gets from the symbiotic ('mutualistic' is the term in the syllabus) relationship; however, there are some relevant points in terms of the essay title.

Carnivores are another trophic level up. They acquire their nutrients from predation and eating other animals. Inside all these animals may be parasites. Internal parasites like worms receive their nutrients by hooking themselves in the small intestine. They are long and flat so there is a very large surface area for digested food to diffuse in.

A logical sequence of events, considering the tertiary trophic level after producers and primary consumers. However, John does not give any examples. You will be expected to give named examples and here the term 'worm' is too vague. John does not consider adaptations to this mode of nutrition – development of sensory systems, teeth, claws, etc. – which would have been easier if an example was used. He does refer to internal parasites but only mentions modification to this mode of life in terms of its flat shape increasing surface area for absorption. He could have mentioned modification of digestive systems and structures to maintain position in the host – hooks and suckers, for instance.

All mammals receive their nutrients from eating. These are then broken down in the gut by enzymes until they are small enough to be absorbed by the villi. Microvilli in the gut provide a huge surface area for absorption.

The first sentence is a GCSE-type statement. John had a chance to mention the specificity of digestive enzymes with examples but did not take it. He needed to give the general features of digestive enzymes – i.e. that they are all hydrolytic. He did not go into enough detail about the modification of the ileum for absorption. The structure of the villi in terms of blood and lymphatic supply, the large number of mitochondria in the epithelial cells and the nature of the microvilli could have been mentioned.

> Marine organisms also obtain nutrients through the same way. <u>Fish eat many plankton and whales are filter feeders.</u> This means they have a sieve ✓ like structure which picks up plankton as it swims around. Many protoctista receive their nutrients by <u>diffusion</u> alone. Fungi are <u>saprophytes</u> and release enzymes ✓ like amylase ∧ which breaks down the organic product outside. The fungi can then absorb the broken down nutrients. An example is a pin mould.

John offers examples of different methods of nutrition. However, all fish do not eat plankton and not all whales are filter feeders as he suggested. He had a chance here to consider endocytosis as a method of obtaining nutrients but he incorrectly refers to diffusion. The term 'saphrophyte' is incorrectly used as the fungi are not plants ('saprobiotic' is the correct term) but this method of nutrition is clearly understood.

> There are many adaptations in obtaining nutrients. One of them is a Venus flytrap. These plants can live in places with low nutrients and ∧ they supplement their diet by enticing flys and insects with their scent. ✓ When a fly lands on it and triggers a hair, the trap closes and enzymes are released.

A carnivorous plant is mentioned (this is not on the syllabus – however, relevant information will be credited if it is at the correct level). In this case he offers no detail as to what nutrient (i.e. nitrogen) is lacking in the environment or why 'eating' flies helps.

> Mosquitoes take their nutrients directly from the blood ✓ of animals. They have a syringe like proboscis which pierces the skin and releases anti-clotting agents.

A final brief reference to mosquitoes. As an ectoparasitic modification this could have been included earlier – it looks like an afterthought but not one that relates structure to function.

Marks awarded for this essay

The marks awarded and the reasons why they are given are explained below.

Scientific content (maximum 16 marks)
6 marks awarded

The general impression is that John has covered a lot of material but at too superficial a level to score in the high band. He hasn't developed many ideas to full 'A' level standard. Enzymes are mentioned a few times but only amylase is named and the importance of enzymes in the digestive process is never considered. Some aspects show an understanding of this area of biology and he has drawn together aspects from a number of different topics.

This will score at the top of the 'poor' band or the bottom of the 'average' band.

There are a few biological errors but not many. John has attempted to consider bacterial involvement and did correctly name the ions involved.

2 Breadth of knowledge (maximum 3 marks)
3 marks awarded

Most of the areas that could be expected from an 'A' level candidate are covered, if only briefly. Examples from all trophic levels are given; photoautotrophs, chemoautotrophs, primary consumers, secondary consumers. Even specialised modes of nutrition such as parasites, symbionts, carnivorous plants are described.

3 Relevance (maximum 3 marks)
3 marks awarded

All of the material offered is relevant. John even repeated or rephrased the title in most paragraphs to focus the examiner on the material he was writing.

4 Quality of language (maximum 3 marks)
3 marks awarded

There is a logical development of ideas throughout the essay (Marks therefore must fall in the 3/2 category) Spelling, punctuation and grammar is consistent with what might be expected of an 'A' level candidate and he uses the correct scientific and technical terms in context.

Total marks 15/25

Practise marking your essays

When you have written an essay it is helpful to swap with a friend and mark each other's work. If that is not possible try to mark your own before you hand it in. Get used to looking for marks and be aware of what the examiner is going to expect.

1 Scientific content

To help you see how many marks you can expect for 'scientific content' put a tick in your essay when you make a point which you are sure is of 'A' level quality – check the specification if you are in doubt about the standard or ask your teacher for advice. Underline anything that is wrong. As you will have written the essay without notes or textbook check areas where you were unsure of your facts. The numbers of ticks and the underlining should give you a rough idea of the band the essay is in. Whole paragraphs with correct GCSE standard work should not get ticks. Be very critical.

❷ Breadth of knowledge

To see what you can expect for 'breadth of knowledge' start a new paragraph whenever you start a new topic. See how many paragraphs you have written. The areas of the specification you have identified at the beginning of your plan ought to give you a rough idea. In some cases not all the headings of your structure will be covered but you should aim to cover at least three of the five.

❸ Relevance

'Relevance' can be subjective and sometimes you will want to use something that is not directly related to the topic to expand your explanation of something that is relevant. You may also want to introduce the topic by putting it into context. Examiners will be very understanding and tolerant if this is the case. When you have written your essay try to put a wiggly line at the side of any passage that does not relate directly to the essay title. Compare the number of wiggly lines with the total content of the essay – you should be able to see whether you have strayed from the point too often.

❹ Quality of language

Your plan will help you organise your ideas logically and you should get a high 'quality of language' mark by considering your language and using the right terms.

- This is a piece of scientific writing so do not use phrases like 'and now I am going to tell you about'. Examiners also get frustrated with phrases like 'the cell thinks it needs more oxygen', because it cannot think, or 'the brain sends messages to the skin', because they are not messages – they are nervous impulses.

- The essay is also intended to be continuous prose, so avoid bullet points or notes unless you are desperate and have run out of time. A few notes at the end of an essay may get you extra credit for scientific content without losing you marks for quality of language; as long as the examiner has enough evidence of your ability.

- Avoid abbreviations like PS for photosynthesis, although ATP and DNA are acceptable.

- Try to spell consistently. Even if you are unsure of the spelling of a word use the same spelling each time you use that word. Better still, if you cannot spell a word use another.

- Use capital letters at the beginning of sentences but do not spread them around at random, and use paragraphs to separate different parts of the essay.

- Don't give human emotions to other organisms – plants don't *know* that it is time to flower; woodlice aren't *happy* in rotten vegetation.

Here are a number of essay titles – try to make plans for as many as you can and then see if you can write some of them.

Instructions

The instructions in the paper will be:

- Write an essay on one of the following topics.
- You should select and use information from different parts of the specification.
- Credit will be given not only for biological content but also for the selection and use of relevant information and for the organisation and presentation of the essay.

You will always have a choice of topic.

Essay titles

How microorganisms can benefit humans

The different ways in which living things obtain their nutrition. **(There is a 'model' essay at the back of the book (p. 95) for this title.)**

The function of mineral ions in animals and plants

The passage of water through plants

The functions of protein

How the structure of cell organelles is related to their function

The function of cell-surface membranes

The effect of temperature on living organisms and the processes which occur in them

How the structure of different cells is related to their function

The ways in which genes and the environment affect the phenotype of an organism

Osmosis and its importance in living organisms

The effect of ecological conditions on the distribution of organisms

The process of diffusion and its importance in living organisms

Relationships between different species of organisms

How different organisms obtain the element nitrogen

The similarities and differences between nervous and hormonal control in animals

The part played by microorganisms in nutrient cycles

Adaptations to living on land

Homeostasis in a mammal

Answers to Questions to try

Chapter 1 Inheritance

Q1 How to score full marks

(a) (i) Blood group B ✓

> **Examiner's comments**
>
> You are not asked for an explanation for this answer. However, from the diagram it is possible to work out all the genotypes and phenotypes. As blood group O is the result of two recessive alleles the child with this blood group must have the genotype I^oI^o. Thus the father with blood group A must have contributed one of the I^o alleles – his genotype must be I^AI^o. The mother must have contributed the other I^o allele. However, the child with blood group B can only have received the I^B allele from its mother, so the mother's genotype must be I^BI^o.

(ii) 1/8 ✓

> **Examiner's comments**
>
> There is a 25% chance of the next child having blood group AB – remember the chance never depends on what has happened before – and a 50% chance of that child being a girl. So the probability of the two events is found by multiplying 50% by 25%.
>
> 50% of 25% = 12.25% or 1/8 or 1 in 8 (*not* 1 to 8 or 1:8 – this means 1 in 9).

(b) (i) $I^o = 1 - (0.19 + 0.06)$

$\qquad = 1 - 0.25$

$\qquad = 0.75$ ✓

> **Examiner's comments**
>
> The total frequency of all alleles of a gene – whether there are two or three or more – is 1. So the frequency of the third allele in this case is 1 minus the sum of the two you have been given.

(ii) Frequency of the allele I^o is 0.75 = q ✓

Thus the frequency of the genotype I^oI^o is q^2

$0.75^2 = 0.56$

$\qquad = 56\%$ ✓

> **Examiner's comments**
>
> If you know the frequency of the allele, the frequency of the homozygous genotype is the square of that value. If you know the frequency of the homozygous genotype (the homozygous recessive is often the only one you can recognise in a question) the frequency of the allele is the square root of that value.

Q2 How to score full marks

(a) Meiosis. ✓ Two different diagrams show chromosomes attached to the spindle. In one the chromosomes appear to be a single structure, in the other a double structure. This suggests that in ✓ this type of division there are two stages. Only meiosis has two stages.

> **Examiner's comments**
>
> Mitosis has a single stage when the replicated chromosome – consisting of a pair of chromatids – separates (anaphase). In meiosis homologous chromosomes separate first and then during a second stage the chromatids separate. It may also be possible to distinguish between the two processes by identifying 'bivalents' (homologous chromosomes linked together), either when they first associate – diagram C – or when they are separating on the spindle – diagram B.

(b) CBDA ✓

> **Examiner's comments**
>
> During the first part of meiosis chromosomes consist of two chromatids. Therefore B and C must be the first two stages.
>
> Homologous chromosomes come together to form bivalents so that chiasmata and cross-over can occur (diagram C) before homologous chromosomes separate (diagram B). So C is the first stage and B is the second.
>
> During the second part of meiosis chromatids separate. Therefore A and D must be the last two stages.
>
> Chromosomes become attached to the spindle (diagram D) before the chromatids separate and move toward the poles of the cell (diagram A). So D is the third stage and A is the fourth stage.

(c) Crossing over swaps pieces of DNA and forms new chromatids. ✓ When the chromatids separate each becomes a chromosome in a gamete. This process therefore increases genetic variation by ✓ producing new combinations of genes on new chromosomes.

Examiner's comments

The question asks you to 'explain' the 'importance' of the events and so although you have to recognise that this is cross-over, you will not get any credit for that fact alone. To get both of the available marks you need first to explain what cross-over creates (i.e. new chromatids) and then give the significance of that in terms of genetic variation.

Chapter 2 Ecology

Q1 How to score full marks

(a) $377 - 224 = 153 \text{ kg ha}^{-1} \text{ year}^{-1}$ ✓

Examiner's comments

'Net' gain takes into account input and loss. Decomposition (F), nitrogen fixation (B) and fertilisers (C) add nitrogen. Uptake by crop (E), leaching (A) and denitrification (D) remove it.

(b) (i) F ✓

Examiner's comments

Nitrifying bacteria have enzymes that oxidise ammonium ions into nitrites or nitrites into nitrates. The bacteria make use of the energy released to synthesise organic chemicals. As nitrification is an oxidation these bacteria need oxygen. When looking at questions, availability of oxygen is sometimes a clue that this type of bacteria are involved.

(ii) B ✓

Examiner's comments

Nitrogen-fixing bacteria contain an enzyme that 'fixes' the element by combining nitrogen gas with hydrogen. The bacteria can then use the ammonium ions created. If the nitrogen-fixing bacteria are found in the root nodules of leguminous plants then those plants can use this source of nitrogen.

(c) (i) Inorganic nitrogen compounds are reduced to form nitrogen gas. ✓

Examiner's comments

Denitrifying bacteria are anaerobic, and live in conditions where oxygen levels are low. They obtain their energy by removing the oxygen from nitrites or nitrates. This process creates gaseous nitrogen, a form that plants cannot use. Whereas nitrifying bacteria produce soluble ions that plants can absorb, denitrifying bacteria produce a gas that they can't.

(ii) Sand is made up of larger particles and therefore there are more air spaces containing oxygen between them. Denitrifying bacteria are anaerobic ✓ and do not survive well when a lot of oxygen is present.

Examiner's comments

Denitrifying bacteria cannot exist in oxygen-rich soils. They are found commonly in waterlogged areas because the water fills the air spaces in the soil. So free-draining, oxygen rich, sandy soil will not be an ideal environment for them. This question expects you to make the link between 'waterlogged' and no oxygen, sandy soil and lots of oxygen.

Q2 How to score full marks

(a) It will go to decomposers. ✓

Examiner's comments

Energy cannot be created or destroyed – it is just converted from one form to another. The chemicals in faeces and urine contain energy which is used by microorganisms (bacteria) and saprotrophs (fungi).

(b) (i) 5.30 MJ day^{-1} ✓

Examiner's comments

Energy taken in – 'consumption' minus the sum of the energy lost – 'respiration' and 'faeces and urine':

$28.41 - (10.14 + 12.97)$.

As an estimate 28 minus about 23 is 5, so the answer given by my calculator (5.30) is in the right area. Figures given are to two decimal places so use the same level of accuracy. Do not forget to quote the units.

(ii) 19% ✓

Examiner's comments

Your previous answer divided by the energy consumed multiplied by 100 to produce a percentage. (Examiners will calculate this using the answer you have given in (i) to ensure that one error will not lose you two marks). So the calculation is $5.30/28.41 \times 100$.

(c) Keeping animals inside means that you can keep them warm and so they lose less energy as heat. ✓

Examiner's comments

There are two major ways that energy can be lost. Mammals lose lots of energy as heat to maintain their temperature (the answer given here). Any energy used in respiration to allow the animal to move is also lost. So if they have to move about less in a barn to search for food, more of the energy in the food they eat can be used to make 'more cow' or 'more sheep' and productivity will be improved.

Chapter 3 Biochemistry of photosynthesis

Q1 How to score full marks

(a) Thylakoid membranes. ✓

> **Examiner's comments**
> 'Chloroplast membranes' would not be enough – you would be expected to know the correct term for a mark. 'Grana' would have been an acceptable alternative answer (this is where the thylakoids are folded to produce the most concentrated area of chlorophyll). Remember, the light-independent reactions occur in the areas surrounding the membranes – the stroma.

(b) (i) Photolysis. ✓

> **Examiner's comments**
> Water is broken down because of the effect of light. So 'photo' – light and 'lysis' – breakdown. It is easy to see where the word comes from.

(ii) They replace the electrons lost by chlorophyll ✓

> **Examiner's comments**
> Both photosystems would involve chlorophyll and when light excites an electron enough to cause it to leave this molecule, free electrons from photolysis will be attracted to take their place.

(iii) It is used to reduce NADP. ✓

> **Examiner's comments**
> Reduced NADP is NADP with hydrogen attached. The formula of this molecule is quite complicated but you can not go wrong if you always refer to it as 'reduced NADP'.

Q2 How to score full marks

(a) (i) ADP and phosphate are joined to make ATP ✓
NADP is reduced. ✓

> **Examiner's comments**
> Light and whole chloroplasts are available so potentially both reactions could occur.

However, there is no carbon dioxide so only the light-dependent reaction will take place. ATP and Reduced NADP are the end products of this reaction.

(ii) Carbon dioxide would react with RuBP to produce GP. ✓ ATP and reduced NADP are used to convert GP to TP and so they would not be available for Stage 3.

> **Examiner's comments**
> To identify the products of the light-independent reaction a tracer of some kind is necessary. In this case radioactive carbon dioxide is used. If all the ATP and Reduced NADP were used up no carbon dioxide would be fixed in the dark.

(b) Organic compounds can be separated using chromatography and matched with known molecules. The radioactive compounds can be identified using autoradiography. ✓

> **Examiner's comments**
> Exposing radioactively sensitive paper to chromatographs will produce a pattern relating to the presence of radioactive molecules. Although a Geiger counter will measure radioactivity it will not clearly show the position on a chromatograph.

(c) In Stage 3 the radioactive carbon is fixed ✓ and during this stage only the stroma are present. ✓

> **Examiner's comments**
> The diagram shows that the chloroplast has been broken down and the membranes have been removed but only after allowing them to produce ATP and Reduced NADP. The light-independent reaction, however, can not occur in the dark unless (as in this case) it is provided with the necessary molecules.

Chapter 4 Biochemistry of respiration

Q1 How to score full marks

(a) Krebs cycle occurs in the mitochondria, glycolysis does not. ✓

> **Examiner's comments**
> Glucose is the substrate for glycolysis, which takes place in the cytoplasm. However, the stem of the question states that the preparation being used in this experiment is of

mitochondria, so it would not contain the enzymes to break down glucose. Pyruvate is the molecule produced by glycolysis, which is absorbed into the mitochondria, but there will be enzymes present to metabolise any of the subsequent molecules created by the link reaction or by Krebs cycle.

(b) Phosphate. ✔

> **Examiner's comments**
>
> Looking at the diagram you can see that there is a substrate added, a source of energy to make ATP. Oxygen is supplied, so the hydrogen carrier system will function. ADP is there too and all the necessary enzymes and carriers from the mitochondria. The only raw material missing is phosphate.

(c) **(i)** ATP is being produced ✔ and oxygen is the hydrogen acceptor. ✔

> **Examiner's comments**
>
> The mitochondria are the site of oxidative phosphorylation. Oxygen is the final acceptor of hydrogen at the end of a series of carriers. Without oxygen none of the other carriers can pass on their hydrogen (or electrons in some cases), so the whole chain of reactions stops. While ATP is being made oxygen is being used up, combining with hydrogen to produce water.

(ii) Another factor apart from ADP is limiting ATP ✔ production.

> **Examiner's comments**
>
> When we look at graphs, we are all used to the curve with a positive slope (the line goes up and to the right) which then flattens off. We know that after the point when it flattens the variable represented on the x axis is no longer the limiting factor. (Think of the graph showing how the rate of photosynthesis changes when intensity of light is increased) This graph shows a similar pattern but it is upside down. A number of things could stop further use of oxygen – further aerobic respiration – such as lack of substrate or phosphate.

Q2 How to score full marks

(a) **(i)** Oxidative phosphorylation. ✔

> **Examiner's comments**
>
> The cristae of the mitochondria are the folds in the inner membrane. The membrane is a phospholipid bilayer with proteins embedded in it. These proteins are the carriers of the hydrogen carrier system and the larger the area of the membrane the more carriers can be present.

(ii) Link reaction *and* Krebs cycle ✔

> **Examiner's comments**
>
> Both stages are needed for a single mark here. Many students would remember that the main function of the Krebs cycle is to break down organic molecules, releasing hydrogen and producing carbon dioxide. Fewer would recall that pyruvate has three carbon atoms per molecule whilst acetyl coenzyme A has two. What happens to the other carbon atom? It is released as carbon dioxide.

(b) They have to respire anaerobically, which means ✔ that a larger number of glucose molecules have to be broken down to produce the same amount of ATP.

> **Examiner's comments**
>
> When water turns to ice it floats and acts as an insulating layer. This stops fish and other aquatic organisms from freezing to death but oxygen diffuses more slowly from the air through solid water. Lack of oxygen in the water will mean that fish have to respire anaerobically. One molecule of glucose will result in the production of 38 ATP molecules by aerobic respiration but only 2 ATPs by anaerobic respiration.

(c) In oxidative phosphorylation a chemical reaction provides the energy but in photosynthesis the energy comes from light. ✔

> **Examiner's comments**
>
> ATP is made during photosynthesis by photophosphorylation. There are a number of similarities – ADP and phosphate are combined; electrons are passed along a number of carriers; the carriers are held in membranes (thylakoids in the chloroplast, cristae in the mitochondria). The major difference in this case is that the electrons are removed from chlorophyll when this molecule is excited by light energy.

Chapter 5 Water transport in plants

Q1 How to score full marks

(a) **(i)** Loss of water from the surface of a plant. ✓

> **Examiner's comments**
> Transpiration is the loss of water vapour. The water changes into a gas within the plant and this diffuses into the air through stomata (pores of variable diameter in leaves) and lenticles (small holes in the corky layer of stems).

(ii) That the amount of water lost is the same as the amount of water taken up. ✓

> **Examiner's comments**
> A potometer measures the amount of water taken up by a plant under different conditions. Although the water may be used for photosynthesis or if the plant is dry at the beginning of the experiment to physically support its cells – making them turgid – it is presumed that water use by the plant will be constant.

(b) The distance the air bubble moves in a set period of time. ✓

> **Examiner's comments**
> If the diameter of the capillary tube is known it is possible to calculate the volume of water taken up (lost) per minute or per hour. However, the plant must be allowed to get used to the conditions before any measurements are taken. It is also important to repeat the experiment a number of times and calculate a mean rate.

(c) To be sure that no air gets into the xylem. ✓

> **Examiner's comments**
> The water in the xylem moves as a continuous column as the molecules of water cohere (stick) to one another. If an air bubble enters the xylem the column will be broken and, although the top of the column will continue to be drawn up the stem, no water will be taken in at the bottom. As this experiment relies on measuring the uptake of water it will not work.

(d) Use radioactive tracers. ✓

> **Examiner's comments**
> It is obviously silly to think that you could put a tree in a potometer but it may be possible to introduce a dye or radioactive chemical which could be measured as it moves up the trunk.

Q2 How to score full marks

(a) Curve approx. the reverse of that for the diameter of the branch with its peak at about 12.00. ✓

> **Examiner's comments**
> The rate of transpiration is affected by humidity, temperature and light, all of which have the greatest effect at mid-day. If wind speed does not change you could expect the highest rate at mid-day and the lowest rates in the evening and early morning.

(b) **(i)** A – cohesion/tension. ✓

> **Examiner's comments**
> Both root pressure and capillarity force water into the xylem cell. These are unlikely to cause it to collapse and do not produce a large enough force to have any affect on the diameter of the cells.

(ii) When the rate of transpiration is at its maximum the water column is under tension, which reduces the diameter of the xylem. ✓

> **Examiner's comments**
> Transpiration reduces the pressure at the top of the xylem vessel as water is removed and the cell acts like a straw. If the sides of a straw were too thin, it would collapse. Even though the walls of xylem are thickened with lignin the tension created inside the cell pulls the wall inward. This, repeated in the thousands of xylem cells, produces a measurable decrease in the diameter of a tree.

Chapter 6 Homeostasis

Q1 How to score full marks

(a) When the body temperature departs from the norm a mechanism is initiated which returns it to the normal temperature. ✓

> **Examiner's comments**
> Although it is possible to define negative feedback without reference to an example, in this case the question expects you to use temperature as an example. The diagram

shows that both higher and lower than normal temperatures will cause a comparable effect – that is, to return the body temperature to the norm – so it would be better to mention them both.

(b) Hypothalamus ✓

> **Examiner's comments**
>
> This structure monitors blood temperature. Do not confuse this with the skin receptors that detect changes in environmental temperature.

(c) Vasodilation of arterioles occurs. This causes ✓ more blood to flow to the surface of the skin where more heat is lost by radiation.

> **Examiner's comments**
>
> Dilation means that the vessel gets wider (not tighter) and therefore more blood gets to the surface of the skin. Remember, some blood does get through the vessel at all times so stress the 'more'. No marks will be given for writing only 'blood gets to the surface of the skin'. Only arterioles in the skin dilate – capillaries don't. Avoid using the term 'vessels' as this would be too vague. Do not forget that at A level you ought to be able to give the process by which heat is lost. A really common mistake is to say that 'blood vessels move to the surface of the skin'.

(d) This gives the body greater control so you do not have to wait for the temperature to fall naturally. ✓

> **Examiner's comments**
>
> Most systems in the body work antagonistically. Think of the muscles – one moves a limb in one direction and the antagonistic muscle moves it in the opposite direction. The same happens here, and the disadvantage of a slow movement in one direction is obvious.

Q2 How to score full marks

(a) If the concentration of glucose in the blood increases, this leads to an increase in the ✓ production of insulin by the pancreas. Insulin becomes attached to protein receptors on the ✓ cells of the liver, causing them to take up more glucose from the blood. It also activates an enzyme within those cells, stimulating the ✓ conversion of glucose to glycogen and thus ✓ lowering blood glucose levels. If blood glucose levels fall the pancreas secretes glucagon. This hormone causes the conversion of glycogen to glucose in the liver and blood glucose levels rise.

> **Examiner's comments**
>
> This question concentrated on the last two phases described in the hints. Both hormones are produced by the pancreas – if you know exactly which cells (α or β cells of the islets of Langerhans) are involved you could name them, but do not guess. Remembering the names glucogenesis or glycogenolysis may be a challenge but unless you know what they mean and the significance of the processes do not waste your time. At A level you ought to know more detail of the effects of the hormones – how does insulin decrease the blood glucose concentration? How does glucagon increase it?

(b) Ultrafiltration occurs between the glomerulus and ✓ the Bowman's capsule. When the blood passes from the renal afferent arteriole into the glomerulus, pressure is produced. This rapidly ✓ removes glucose from the blood as it is such a small molecule. Reabsorption occurs in the ✓ proximal tubule where protein carriers speed up ✓ the movement of glucose from the nephron by facilitated diffusion. ✓

> **Examiner's comments**
>
> Examiners expect terms like 'ultrafiltration' to be used and no marks are given for expressions like 'pressure filtration'. Remember all glucose molecules are reabsorbed from the proximal convoluted tubule and therefore this cannot be by diffusion alone. Active transport of the glucose from the lining cells – epithelium – into the blood maintains the concentration gradient until all the glucose has been removed from the filtrate.
>
> Knowledge of the correct names of parts of the kidney is also important, so you can confidently describe where things happen. The proximal convoluted tubule is often referred to as the first convoluted tubule and the region after the loop of Henle, the distal convoluted tubule, as the second convoluted tubule; the Bowman's capsule is referred to as the renal capsule. These alternative names are quite acceptable.

Chapter 7 The kidney

Q1 How to score full marks

(a) As there is a high pressure in the glomerular ✓ capillaries, particles are forced out of the blood into the renal capsule. This is known as ultrafiltration. Pores in the basement membrane ✓

allow only small molecules such as urea and glucose through, but larger molecules like protein will not leave the blood. In the first convoluted tubule all the glucose is reabsorbed. This occurs by facilitated diffusion down a concentration gradient and then by active transport.

Examiner's comments

The filter in the kidney is made of three layers but the cell layer of the renal capsule and the single cell layer of the capillary are stuck to a non-cellular layer called the basement membrane. The cells do not join well and therefore there are lots of spaces between them. The only barrier is thus the basement membrane and this is the actual filter.

So that all the glucose is removed from the tubule and returned to the blood the glucose in the lining cells of the first convoluted tubule is passed out by active transport. This ensures that a diffusion gradient is maintained and 100% of the glucose is removed from the tubule.

(b) Water is reabsorbed from the descending limb of the loop of Henle by osmosis. Diffusion and active transport of sodium chloride from the ascending limb reduces the concentration of solutes in that limb and creates a concentration gradient in the medulla. Water is reabsorbed from the collecting duct as there is a higher concentration of solutes in the medulla.

Examiner's comments

Remember, the descending limb of the loop of Henle is permeable to water and therefore water is reabsorbed. This makes the filtrate more concentrated. The ascending limb, however, is impermeable to water but sodium and chloride ions pass out. This makes the filtrate less concentrated and creates a diffusion gradient across the medulla. As the filtrate passes through the collecting duct it is always in contact with a solution outside that is stronger – that has a lower water potential. So water moves out of the filtrate, which means that the urine becomes more concentrated.

(c) If there is protein in the urine there must be less in the plasma of the blood. This results in the blood having a higher water potential and so less tissue fluid is recovered.

Examiner's comments

At the venous end of the capillary the water potential is normally lower than that in the

tissues, which causes water to be returned to the blood. The water potential is reduced by plasma proteins and so any condition – such as damage to the kidney or malnutrition – which reduces the concentration of protein in the blood will cause water retention.

(d) Transplanted cells have different antigens to those of the body cells. These are recognised by T cells, which attach to the antigen and destroy the foreign cell.

Examiner's comments

Following recognition of the non-self antigen, T cells (lymphocytes) become sensitised and divide by mitosis – they clone – each of the resulting cells responds to this particular antigen and destroys the cell it is attached to. This is known as cellular immunity.

Note that this part of the question links the kidney with immunity – an AS topic.

Q2 How to score full marks

(a) Renal artery ✓

Examiner's comments

There will still be recall questions in A2. You will be expected to know the names of the major blood vessels that serve the kidney, liver and heart, so learn them.

(b) Ultrafiltration ✓

Examiner's comments

Filtration under pressure. There is no time to allow the small particles of the blood to pass through a filter. The vessel that leads into the glomerulus is wider than the vessel leaving and therefore a higher pressure is created – enough to speed up the filtration. Should the blood pressure be too high the filter can be damaged, resulting in large molecules from the blood – such as proteins – being found in the urine.

(c) C ✓

Examiner's comments

Glucose is the major respiratory substrate – it is the main chemical in the body that cells use as an energy source – but it is a small molecule. Therefore it will pass through the filter in the renal capsule. The cells of the first convoluted tubule are modified for the rapid and complete reabsorption of glucose. They have many microvilli, increasing surface area

> for rapid diffusion, and many mitochondria, producing ATP for the active transport needed to ensure complete reabsorption.

(d) The longer the loop the higher the concentration of salts in the medulla. This results in a lower water potential, so more water can be removed from the collecting duct and the urine can be more concentrated than sea water.

> **Examiner's comments**
>
> The concentration of the filtrate at the beginning and end of the loop of Henle is very similar because, although some water has left the nephron, so has quite a lot of sodium chloride. These ions remain in solution in the medulla, surrounding the nephron. The concentration of this solution increases through the medulla, from the cortex to the pelvis regions. The medulla always has a lower water potential than the filtrate in the collecting duct, so water is always reabsorbed by osmosis. The longer the loop, the greater the concentration gradient, the more water can be reabsorbed from the collecting duct and the more concentrated the urine will become.

Chapter 8 Digestion

Q1 How to score full marks

(a) The microvilli have been broken down. ✓

> **Examiner's comments**
>
> 'Describe' can be read as 'write what you can see'. You need not add the effect that this change will have – that would be the answer to the question 'explain'

(b) (i) Enzymes situated in the membrane of the microvilli are involved in the digestion of dipeptides. Less membrane would mean fewer enzymes and therefore the rate of dipeptide digestion would be reduced.

> **Examiner's comments**
>
> Dipeptidase enzymes (and also enzymes that digest disaccharides such as maltase, sucrase and lactase) are attached to the epithelial membrane. As the area of this membrane is reduced by these bacteria so will the number of the enzymes present and the effectiveness of the digestive process.

(ii) This would also be reduced, as the rate of absorption is directly proportional to the surface area of the membrane.

> **Examiner's comments**
>
> The products of dipeptide digestion are amino acids, which are absorbed by facilitated diffusion. Not only will there be a reduced surface area (which affects diffusion rates) but also there will be fewer protein carriers. Either of these things will reduce absorption and either would be an acceptable answer.

(c) Membranes are made up of phospholipid and protein but complex carbohydrates on the surface of this membrane act as a barrier, keeping the lipid and protein digesting enzymes away from the membrane. From the diagram it can be seen that *E. coli* infection removes that protection.

> **Examiner's comments**
>
> There are many differences in the two diagrams – two mitochondria instead of one, a different-shaped golgi apparatus, more vacuoles – but you need to read the question carefully and visualise the position of the structures in the body. The pancreas secretes enzymes, which travel down the pancreatic duct into the small intestine. They will therefore come into contact with the outside of these cells. Apart from the change in surface area the only other visible difference at the surface of the cell is the presence of complex polysaccharide molecules embedded in the membrane. From that point it is a small step to link the specificity of the enzymes mentioned in the question to the fact that they will not be able to digest the polysaccharides but could digest the components of the membrane should they come in contact with it.

Q2 How to score full marks

(a) Circular muscles contract and longitudinal muscles relax above the food, which reduces the diameter of the tube and pushes food down the intestine.

> **Examiner's comments**
>
> This process is called peristalsis. As the circular muscle contracts the tube gets thinner – this decrease in the diameter of the lumen squashes food. This happens above the food. Alternatively, as the longitudinal muscle contracts the tube gets wider – this increase in the diameter of the lumen opens the tube to receive food. This happens below the food. Only the first action moves food and the question asks for a description of the action of both muscle layers.

(b)

Stimulus that triggers digestion	Effect	Digestive juice secreted
Food in the mouth ✓		
	Gastrin ✓	
		Bile ✓
	Secretin ✓	

Examiner's comments

Gastrin, secretin, pancreozymin and cholecystokinin are the only four hormones involved in the control of secretions. You must know what each does and what causes each to be released.

Chapter 9 Nervous system

Q1 How to score full marks

(a) **(i)** An isotonic solution ✓

Examiner's comments

The term is not necessary but the idea must be clear. There must be the same concentration in the solution suspending the cell as there is inside the cell.

(ii) So that it has the same water potential as the inside of the cell. ✓

Examiner's comments

The neurone is a cell and if it is placed in an environment with a different water potential it will either absorb too much or lose too much water. The cell will therefore either collapse or burst, and it will not be possible to carry out any experiment to see how this cell works.

(b) They are below the threshold value. ✓

Examiner's comments

It is important to remember that you need the correct intensity of stimulus before an action potential is produced.

(c) **(i)** **1** They are of the same size. ✓

   2 They travel at the same speed. ✓

Examiner's comments

Although you have no evidence for the second answer remember your theory. The question does not ask you to used evidence from the data provided; all action potentials in the same neurone travel at the same speed and are of the same intensity.

(ii) Higher intensity of stimulus produces a greater frequency of action potential. ✓

Examiner's comments

The stronger the stimulus, the more rapidly one impulse follows another down a neurone. So 5 impulses in a second may represent a weak stimulus; 10 a second a medium level of stimulus and 20 a second a strong stimulus. The number of impulses a second is the frequency.

(d) Sodium channel proteins open and more sodium ions move into the neurone. The potential difference across the membrane changes from negative (–70 mV) to positive (+40 mV). Sodium channels close and potassium channels open. More potassium ions leave the neurone and the potential difference of the membrane changes from positive to negative, returning it to a resting potential.

Examiner's comments

Sodium ions move through channel proteins (these are often called ion gates). Although you may not get any extra credit for knowing the actual potential difference of a cell when at rest and during the depolarisation stage (action potential) it may be worth giving them to be sure of credit. Terms like 'negative' and 'positive' in themselves may be thought too vague.

Q2 How to score full marks

(a) Curare blocks the receptor molecule so that the neurotransmitter cannot stimulate the muscle. Therefore the muscle will not contract. ✓

Examiner's comments

You need not know anything about this molecule. Compare the 'normal' diagram with the diagram showing the curare effect and it is clear that the site on the membrane of the muscle is being blocked. You can therefore deduce that the muscle will not receive the information to contract.

(b) Organophosphates combine with the enzyme acetylcholinesterase, which normally breaks down the neurotransmitter. The neurotransmitter continues to stimulate the muscle, which contracts and continues to do so. ✓ ✓

> **Examiner's comments**
>
> Just as in a synapse the neurotransmitter must not be allowed to remain in contact with the receptor because while it does action potentials will be generated. In the case of the nerve–muscle junction the effect of this would be continued contraction of the muscle. As the question asks you to use information in the diagram you ought to give the name of the enzyme.

(c) Botulin toxin stops the fusion of vesicles with the membrane of the neurone and therefore the release of the neurotransmitter. Therefore the muscle does not contract. ✓ ✓

> **Examiner's comments**
>
> Without the stimulus of the neurotransmitter the connection between the nervous system and the muscle is broken. Despite nervous impulses being sent to the muscle it will not contract – you are paralysed.

Chapter 10 Transport of respiratory gases

Q1 How to score full marks

(a) **(i)** The unloading tension would increase. ✓

> **Examiner's comments**
>
> The higher the concentration of carbon dioxide, the more of it will combine with haemoglobin and therefore the less oxygen will be carried. This causes the dissociation curve to move to the right – the Bohr shift.

(ii) Haemoglobin would unload oxygen at a higher partial pressure and so would give up more oxygen to actively respiring tissues. ✓ ✓

> **Examiner's comments**
>
> This is important during exercise. As the level of oxygen in the muscle falls, to maintain activity the muscle either has to obtain more oxygen from haemoglobin in the blood or has to respire anaerobically. Anaerobic respiration leads to the production of a waste product – lactate (lactic acid) – which stops the muscle working. The Bohr shift ensures that an active

muscle – with high concentrations of carbon dioxide (the waste product of aerobic respiration) – will be able to obtain oxygen from haemoglobin at relatively high partial pressures of oxygen (i.e. when there is still some oxygen available in the muscle). Therefore the muscle need not respire anaerobically and can continue to work for longer.

(b) **(i)**

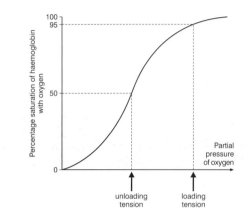

(ii) As there is little oxygen in the burrow, this form of haemoglobin must have a high affinity for oxygen so that it becomes saturated at a lower partial pressure than human haemoglobin. ✓ ✓

> **Examiner's comments**
>
> There are lots of other examples of animals that live in surroundings with low oxygen partial pressures – llamas that live at high altitude, worms that live in mud, for example. All these creatures will have haemoglobin with dissociation curves to the left of that shown here.

Q2 How to score full marks

(a) The cells of the small intestine are active so there is a low partial pressure of oxygen. The graph shows that haemoglobin has a low affinity for oxygen under these conditions. Haemoglobin releases oxygen, which diffuses into the cells of the intestinal wall because there is less oxygen there. ✓ ✓

> **Examiner's comments**
>
> As the cells of the wall of the intestine are active, they will be using oxygen, lowering the oxygen partial pressure and creating the conditions under which haemoglobin releases oxygen.

(b) There is very little oxygen in the small intestine where the parasite lives. Parasitic haemoglobin will be fully saturated at these low partial pressures of oxygen.

11 Synoptic questions – essay writing

Title: The different ways in which living things obtain their nutrition

Plan:

Prokaryotes	Diffusion, etc. Endocytosis Symbiosis Chemosynthesis
Protoctista	Parasitic
Fungi	Saprobiotic
Plants	Photosynthesis
Animals	Heterotrophic

Essay:

Living organisms are classified into 5 kingdoms. This artificial grouping relates not only to their evolutionary relationships and to their morphological similarities but also to the way in which they obtain their nutrients. All organisms need a supply of organic and inorganic molecules from the environment. Carbohydrates, lipids, proteins, minerals and water are all necessary to maintain the structures and to supply energy for the chemical processes occurring in all organisms. However, the size and complexity of each will influence the method used.

Prokaryotes like bacteria are small unicellular organisms and due to their relatively high surface area to volume ratio are able to absorb soluble raw materials from their surroundings by diffusion, facilitated diffusion or active transport. Endocytosis can also be involved. In this case cell membranes fold around an area outside the cell, containing small organic particles. The membrane pinches together and traps the material in a food vacuole. Unicellular protoctista are able to do the same. Once inside the cell lysosomes will fuse with the food vacuole and the hydrolytic enzymes they contain reduces the contents of the food vacuole into small soluble molecules, which can then diffuse into the cytoplasm.

Some bacteria, such as those involved in the nitrogen cycle, are chemosynthetic and use their ability to promote chemical reactions to release energy, which they use to synthesise organic molecules from simpler inorganic molecules. Nitrifying bacteria do this. Another similar method of obtaining nutrients is photosynthesis. Some bacteria and all members of the plant kingdom use light as the energy to make organic molecules. Both chemosynthesis and photosynthesis are known as autotrophic nutrition.

Plants are modified to obtain light – specialised organs called leaves have massive surface areas, and contain a green photosynthetic pigment called chlorophyll within organelles; chloroplasts. Inorganic raw materials like carbon dioxide diffuse into the leaf through pores called stomata, whilst water moves into the leaf in vessels within xylem tissue by means of the transpiration stream and the cohesion and adhesion of water molecules. Photosynthesis itself is a process involving two stages. One, the light-dependent stage, begins when light stimulates chlorophyll to emit an electron. This passes through a number of carriers to generate ATP. The positively charged chlorophyll molecule can accept an electron from water, which causes photolysis, releasing a proton (and oxygen as a waste product). The proton is accepted by NADP, which becomes reduced NADP. The second stage, the light independent stage, involves carbon dioxide being combined with ribulose bisphosphate. The resultant 3-carbon molecule, GP, is reduced to TP (a three-carbon sugar) by hydrogen donated by reduced NADP, whilst the energy and phosphate for the reaction comes from ATP. Both molecules are created in the light-dependent stage. From the molecules described in this process, the plant is able to produce more complex carbohydrates, amino acids, fatty acids and glycerol – all the organic molecules it requires.

Fungi lack the photosensitive pigments needed to convert solar energy into chemical energy, instead they secrete digestive enzymes from their cells. These extracellular enzymes, carbohydrases, lipases and proteases, create small soluble molecules, which can diffuse across the cell membrane along a concentration gradient and into the cell. Fungi can be classified as saprobiotic if they use dead organic sources or parasitic if their substrate is living.

Many members of the kingdoms prokaryotae, protoctista and animalia are parasitic, living inside the bodies of their hosts. Many have adapted to live in the gut, where there is a plentiful supply of nutrients already in a form that can be absorbed. Gut parasites like tapeworm have no digestive systems compared with their free-living relatives, and their flat shape gives them a massive surface area to volume ratio to absorb soluble nutrients through their surface. They also have to develop a method of remaining in one place. The tapeworm has hooks and suckers, which it uses to attach itself to the gut wall. Protoctistans may live in the blood; *Plasmodium*, the organism responsible for the disease malaria, penetrates red blood cells and breaks down their contents.

All animals are multicellular and need to obtain their organic nutrients in a complex form. Unlike plants, they are unable to synthesise these molecules. Such a method of nutrition is called heterotrophic. Most stable organic molecules are either structural or storage materials and as such are large and insoluble. To absorb and transport them they need to break them down into small soluble molecules, and this is achieved by digestion. Many animals have long internal tubes or digestive systems into which digestive enzymes are secreted in sequence. Proteins, for example, are acted upon initially by endopeptidases, which break the long chains of amino acids into small chains. These are then hydrolysed further when the peptide bonds at the end of the chains are broken by exopeptidases, forming amino acids or dipeptides. Final digestion is achieved by peptidases attached to the membranes of the small intestine.

Absorption occurs in the ileum. This is a long twisted tube, the wall of which is further convoluted. On the wall are a series of finger-like projections, villi, containing capillaries and lymph vessels into which the products of digestion diffuse. The membrane of the epithelial cells of the villi also has a number of projections, microvilli. These adaptations give this area a massive surface area and makes absorption of the products of digestion very efficient.

Bacteria that exist in the gut of herbivores have a symbiotic relationship with their hosts. They produce an extracellular enzyme, cellulase, which digests cellulose. This molecule, which makes up the walls of plant cells, is the most common form of carbohydrate eaten by herbivores but without the ability to synthesise cellulase the herbivore is unable to make use of it. The bacteria benefit from the relationship by using the glucose produced by this digestion in their growth. They join nitrogenous compounds to the carbon skeleton, producing bacterial protein.

Herbivores benefit from this relationship as they can obtain protein, a molecule deficient in their diet, by breaking down the bacteria.

Evolution has selected features that have ensured the survival of the organisms living today. Needing nutrients for growth and respiration has meant that the efficiency of these mechanisms has been a strong factor in this evolution.

Why this scores high marks

S – Scientific content [16 marks]:
A great deal of material is given, all at 'A' level standard (note, however, that many things have not been included, such as reference to other digestive enzymes) and there are no major errors.

Possible mistakes:
- Lack of detail.
- Lack of knowledge.

R – Relevance [3 marks]:
Even within the introduction and conclusion reference is made to the essay's title and there is little irrelevant material.

Possible mistakes:
- Writing about anything not related to the title (nutrition) – e.g. life cycle of a parasite.

B – Balance [3 marks]:
A number of areas of the syllabus are considered.

Possible mistakes:
- Concentrating on only one area of the specification – e.g. photosynthesis.

Q – Quality of language [3 marks]:
The essay reads logically with good use of the correct technical terms.

Each paragraph leads on from the last, giving new examples of methods of nutrition. Spelling, punctuation and grammar are acceptable.

Possible mistakes:
- Using technical terms in the wrong place.
- No beginning, middle or end to the essay.
- Common words spelt incorrectly.

Total: 25/25